D1254212

N. H. TECHNICAL COLLEGE AT NASHUA
TS192.M34 1984   c2
Winter, John   Maintenance (sic) manageme

3 0426 0000 5032 9

NASHUA COMMUNITY COLLEGE

MAINTAINANCE MANAGEMENT
FOR QUALITY PRODUCTION

John L. Winter
Editor

Richard S. Zakrzewski
Editor

Robert E. King
Assistant
Manager

Published by

Society of Manufacturing Engineers
Publications/Marketing Services Division
One SME Drive
P.O. Box 930
Dearborn, Michigan 48121

NASHUA COMMUNITY COLLEGE

MAINTENANCE MANAGEMENT
FOR QUALITY PRODUCTION

Copyright© 1984
Society of Manufacturing Engineers
Dearborn, Michigan 48121

First Edition

First Printing

All rights reserved including those of translation. This book, or parts thereof, may not be reproduced in any form without permission of the copyright owners. The Society does not by publication of data in this book, ensure to anyone the use of such data against liability of any kind, including infringement of any patent. Publication of any data in this book does not constitute a recommendation of any patent or proprietary right that may be involv

Library of Congress Catalog Card Number: 83-051535

International Standard Book Number: 0-87263-139-7

Manufactured in the United States of America

TS
192
.m34
1984

SME wishes to acknowledge and express its appreciation to the following contributors for supplying the various articles reprinted within the contents of this book. Appreciation is also extended to the authors of papers presented at SME conferences or programs as well as to the authors who generously allowed publication of their private work.

Automotive Industries
Chilton Company
Chilton Way
Radnor, Pennsylvania 19089

J. Charles Berggren
Monsanto Company
P.O. Box 12830
Pennsacola, Florida 32575

Bruel and Kjaer Instruments, Inc.
185 Forest Street
Marlborough, Massachusetts 01752

Commline
241 Holbrook Drive
P.O. Box 11000
Wheeling, Illinois 60090

Dun's Business Monthly
875 Third Avenue
New York, New York 10022

The Institution of Mechanical Engineers
P.O. Box 24
Northgate Avenue
Bury St. Edmunds
Suffolk IP32 6BW
England

Instrument Society of America
67 Alexander Drive
P.O. Box 12277
Research Triangle Park,
North Carolina 27709

Iron Age
Chilton Company
Radnor, Pennsylvania 19089

Paul Juul
Paul Juul and Associates, Inc.
808 106th Avenue N. W.
Suite 208
Bellevue, Washington 98004

Warren Lee
3556 Normount Road
Oceanside, California 92056

William T. Lesner
Cadillac Motor Industrial Engineering
2860 Clark Avenue
Detroit, Michigan 48232

Richard Murray
Cadillac Motor Car Division
12200 Middlebelt Road
Livonia, Michigan 48150

National Petroleum Refiners Association
Suite 1000
1899 L Street N. W.
Washington, DC 20036

National Safety News
National Safety Council
444 North Michigan Avenue
Chicago, Illinois 60611

Power
McGraw-Hill Publications Company
1221 Avenue of the Americas
New York, New York 10020

Production Engineering
Penton/IPC
Penton Plaza
Cleveland, Ohio 44114

Technical Association of the Pulp and Paper Industry
Technology Park
P.O. Box 105113
Atlanta, Georgia 30348

Terotechnica
Elsevier Scientific Publishing Company
P.O. Box 211
1000 AE
Amsterdam, The Netherlands

PREFACE

As productivity in manufacturing seems to be progressively approaching its growth limit, maintenance has become of of the last cost savings frontiers of management.

As the article from *Dun's Review* (which is reprinted in this volume) points out: "to many executives, maintenance is little more than janitors sweeping floors and technicians changing nuts and bolts." If that is the case, and many maintenance management professionals will tell you that it is, then the challenges presented by that cost savings frontier are truly significant.

For most companies, the cost of avoiding maintenance is enormous. This cost includes labor and material for unplanned equipment rebuild and repair, lost production resulting from unscheduled repairs, and the reworking of products.

In some cases, a poor quality product may be delivered to the customer.

Unfortunately, in the tug of war between production and maintenance, maintenance has consistently proved second best. If industry in the United States is to remain competitive in the world market, this trend must be reversed and the real cost of avoiding maintenance significantly reduced.

To approach this problem, management's attitude about maintenance must change.

Management must understand that there is a problem before anyone can control maintenance costs. This means showing management that avoiding maintenance significantly contributes to waste and the rise in manufacturing costs. Management must understand that effective maintenance practices not only improve uptime, quality, and utilization, as well as reduce repair, labor, and material costs, but prepare industry for the future as well.

Effective maintenance practices prepare for the future by facilitating the implementation of high technology. Existing equipment must be in top condition to accept high technology modifications. Suitable maintenance plans must be in place to get the full benefits from new high technology machines and processes.

Once the impact of effective maintenance practices is fully realized, the opportunities to reduce the cost of manufacturing are endless.

Cost effective maintenance practices are not new ideas. What is new, is the application of these practices. It is the purpose of this collection of articles and papers to provide a reminder of the many maintenance management tools and to assist in the development of new applications.

We wish to thank all of the companies, organizations, publishers and authors who gave permission to have their articles reprinted in this volume. Thanks also to the Publications/Marketing Services staff at SME for their assistance in the research and development required in making this book possible.

John L. Winter
John Deere Component Works

Richard S. Zakrzewski
The Timken Company

ABOUT THE EDITORS

Richard S. Zakrzewski has held the position of Supervisor—Project Management for The Timken Company for the past three years. A veteran of 19 years at The Timken Company, Mr. Zakrzewski is Cochairperson of the SME Maintenance Management Council.

Mr. Zakrzewski holds a bachelor of science degree in Mechanical Engineering from Ohio University and an M.B.A. from the University of Akron. He is a member of the Society of Manufacturing Engineers and the American Society of Mechanical Engineers.

John L. Winter is the Supervisor of Maintenance Engineering of John Deere Component Works in Waterloo, Iowa. He is also cochairperson of the SME Maintenance Management Council.

Mr. Winter has nine years experience with John Deere including work as an electrical maintenance engineer. Previously, he was employed in the Dubuque, Iowa school system as an instructor in industrial education.

Mr. Winter received his B. A. in industrial education and has done post-graduate work in electrical engineering, computer science and vocational education.

SME

The informative volumes of the Manufacturing Update Series are part of the Society of Manufacturing Engineers' many faceted effort to provide the latest information and developments in engineering.

Technology is constantly evolving. To be successful, today's engineers must keep pace with the torrent of information that appears each day. To meet this need, SME provides, in addition to the Manufacturing Update Series, many opportunities in continuing education for its members.

These opportunities include:
- Monthly meetings through three associations and their more than 270 chapters which provide a forum for member participation and involvement.

- Educational programs including seminars, clinics, programmed learning courses, as well as videotapes and films.

- Conferences and expositions which enable engineers and managers to examine the latest manufacturing concepts and technology.

- Publications including the periodicals *Manufacturing Engineering, Robotics Today,* and *CAD/CAM Technology,* the *SME Newsletter,* the *Technical Digest, Journal of Manufacturing Systems,* and a wide variety of text and reference books covering everything from the basics to manufacturing trends.

- The SME Manufacturing Engineering Certification Institute formally recognizes manufacturing engineers and technologists for their technical expertise and knowledge acquired through experience and education.

- The Manufacturing Engineering Education Foundation was created by SME to improve productivity through education. The foundation provides financial support for equipment development, laboratory instruction, fellowships, library expansion, and research.

- A database, accessible through SME containing Technical Papers and publication articles in abstracted form. The Information on Technology In Manufacturing Engineering database is only one of several accessible through the Society.

SME is an international organization with more than 71,000 members in 65 countries worldwide. The Society is a forum for engineers and managers to share ideas, information, and accomplishments.

The Society works continuously with organizations such as the American National Standards Institute, the International Organization for Standardization, and others, to establish and maintain the highest professional standards.

As a leader among professional societies, SME assesses industry trends, then interprets and disseminates the information. SME members have discovered that their membership broadens their knowledge and experience throughout their careers. The Society is truly industry's partner in productivity.

MANUFACTURING UPDATE SERIES

Published by the Society of Manufacturing Engineers, the Manufacturing Update Series provides significant, up-to-date information on a variety of topics relating to the manufacturing industry. This series is intended for engineers working in the field, technical and research libraries, and also as reference material for educational institutions.

The information contained in this volume doesn't stop at merely providing the basic data to solve practical shop problems. It also can provide the fundamental concepts for engineers who are reviewing a subject for the first time to discover the state-of-the-art before undertaking new research or application. Each volume of this series is a gathering of journal articles, technical papers and reports that have been reprinted with expressed permission from the various authors, publishers or companies identified within the book. SME technical committees, educators, engineers, and managers working within industry, are responsible for the selection of material in this series.

We sincerely hope that the information collected in this publication will be of value to you and your company. If you feel there is a shortage of technical information on a specific manufacturing area, please let us know. Send your thoughts to the Manager of Educational Resources, Marketing Services Department at SME. Your request will be considered for possible publication by SME—the leader in disseminating and publishing technical information for the engineer.

TABLE OF CONTENTS

_____ CHAPTERS _____

3   MAINTENANCE METHODS AND PRACTICES

4   COMPUTERS IN MAINTENANCE MANAGEMENT

5 THE JAPANESE APPROACH TO MAINTENANCE

CHAPTER 1

PRODUCTIVITY OF MAINTENANCE WORKFORCE

Reprinted from *Maintenance Management International* (formerly *Terotechnica*), Volume 2, 1981 pp. 205-210, Elsevier Scientific Publishing Company, P.O. Box 211, 1000AE Amsterdam, The Netherlands

INCREASING PRODUCTIVITY BY IMPROVED MAINTENANCE MANAGEMENT

Lawrence Mann, Jr.

Industrial Engineering Department, Louisiana State University, Baton Rouge, LA 10803 (U.S.A.)

(Received July 10, 1980; Accepted December 4, 1980)

Abstract

Maintenance cost is rising faster than other costs attendant to operating a process plant. During 1978, the last year that we have complete figures for, nineteen major companies spent 1.1 billion dollars on maintenance, an increase of 15% over the previous year. A number of identifiable variables affect this cost. In the age of ever increasing accountability it is necessary to isolate maintenance costs and apply productivity indexes to those activities. The indexes must be consistent, relevant and proven. The types of indexes suggested in this paper include those which measure planning, measure control of the work force, and monitor the entire maintenance system. The article also discusses some interpretations given to various productivity indices for maintenance.

INTRODUCTION

Plants have been paying more for maintenance during the past few years than previously. A decade ago the cost of maintenance was so insignificant that management gave little surveillance to this expense. This entire picture is changing. Recently Chemical Week Magazine [1] reported that the cost of maintaining a plant in the chemical process industries set another record in 1978 — the tab for 39 major companies climbed 15% to 1.1 billion dollars. Although costs keep rising, the data show that as a percentage of asset value the maintenance bill remains fairly stable. Twenty selected chemical plants spent 3.8% of the estimated 88.1 billion dollars replacement cost of their assets on maintenance. A comparable group in a survey one year ago spent 4% of their estimated replacement cost of 77.8 billion dollars on maintenance.

Even though the 15% increase in 1978 was less than the 20% climb experienced in 1974 (Fig. 1), the problem is one of increasing

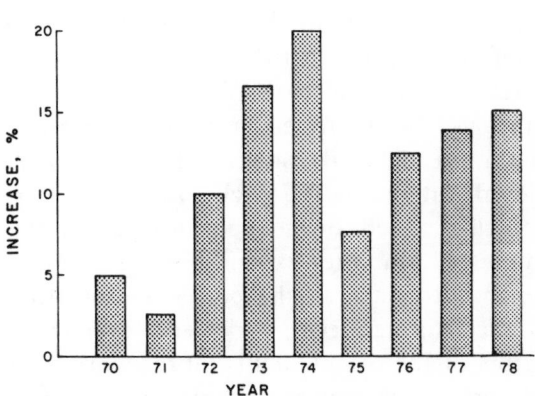

Fig. 1. Maintenance cost trends.

importance and deserves the attention of all maintenance managers.

Maintenance expenditures are considerably influenced by inflation although they are not subject to the large swings often seen in construction costs. However, despite changes in the overall economic picture, we must maintain equipment if we are going to operate a plant.

MAINTENANCE ORGANIZATIONS MUST BE COST-CENTER ORIENTED

To monitor productivity maintenance organizations must be divided into components. Most maintenance organizations contain separate functions for the individual craftsmen who perform the work and for the administrative staff, which includes first line supervision, planners, schedulers, expediters, and others. In any overall productivity monitoring system, the various levels of plant management above those levels mentioned above must be considered, and each level in the organization must have its own unique productivity indicator to determine the performance of those elements it controls.

For instance, the first line supervisor is interested in the ability of the individual craftsmen under him to perform the work that has been assigned to them. He is interested both in whether or not they finish within the allocated time using the resources allocated for the project and in the total idle time attributed to his group. Middle management is interested in the performance of the work force on a weekly basis. Middle management may also be interested in such indicators as the ratio of work orders scheduled to be closed out during the week to those that are actually closed out. Upper management is interested in the expenditures for maintenance to date, in the percentage overtime worked during recent work periods, and indicators of the overall ability of the work force to perform as required. Typical productivity

TABLE 1

Typical productivity ranges for various maintenance situation *.

Maintenance situation	Productivity-range (Percent)
No formal planning-scheduling. Minimal work order system. No measurement.	30– 50
Formal planning-scheduling. Functioning work order system. No measurement.	50– 70
Formal planning-scheduling coordinated with measurement system. Functioning work order and control systems.	70– 90
Formal planning-scheduling-complete systems-complete measurement-incentive system.	90–120

* Productivity is not the percent of time craftsmen are busy, but is a calculated value based on direct work and other appropriate factors.

figures for plants with varying maintenance situations are shown in Table 1.

INDICES

Necessary to any productivity monitoring system is the use of a number of indicators to reflect accurately the ability of the work force to perform as required and to gage the efficiency of this performance. Most organizations find that a number of indexes are necessary to reflect completely the status of maintenance. Those indexes must be used consistently and regularly, and they must be applied to the work force for a number of years to recognize the meaning of trends. Trends are more important than the absolute values of the indexes, which may include sampling errors. Paul D. Tomlingson, a management consultant with Industrial Maintenance Management, Denver, Colorado, reports the effects of such trends, which are illustrated in Table 2. If the index measures are frequently changed, they cannot be validated, which must be done prior to determining whether or not to use them.

It does not matter what indicators are

TABLE 2

Measurement factors

	1973	1974	1975
Worker productivity (percent)	32	37	43
Productive hours per man per year	576	666	774
Cost per productive hour	$16.12	$11.42	$9.21
Percent of materials drawn from warehouse	31	50	63
Percent of foreman's time spent on supervision	37	47	62
Units of product per maintenance dollar	1.51	1.70	1.89
Number of hourly employees	123	110	102
Total maintenance cost	$3.750.000	$3.921.050	$4.017.521

used — they must be tested to determine whether changes in the maintenance work force are adequately reflected in changes in the plants maintenance needs. Nothing makes maintenance more susceptible to criticism than to have widely varying numbers of employees, materials, and equipment available and to have no appreciable change noticed in the degree of downtime. In such a case management would rightly conclude that this probably indicates that the plant is being overmaintained. In testing indicators it would be well to go back through the records for, say, three years and to determine whether the new set of indicators would adequately reflect the changes in maintenance or the changes in production required of the plant. Then one may ascertain whether the data prove that the indicators do actually reflect the change desired by management, and thus may be used as adequate measures of productivity.

From the standpoint of plant maintenance, productivity is reflected in the ability of the direct labor force to adhere to some predetermined standard of activity dictated by management. In addition to the direct labor indicators, a measure of the productivity of the administrative group of the maintenance unit is necessary, since that group plans and schedules the activities and has the opportunity to use the work force either efficiently or inefficiently.

In maintenance terms, productivity is the ratio between some measured aspect of maintenance activity and some standard for what resources should be consumed to complete that task. The many factors that measure the various aspects of maintenance can all be related to some acceptable base, standard, or objective. In this way, the indicators can show the degree to which the actual relates to the theoretical, and as such, are measures of productivity. Just as one never actually measures temperature (but only the effect of temperature on, say, mercury or some other metal), so, one does not measure productivity — one only relates a change in productivity to some variable.

TYPES OF INDICATORS

Indicators of productivity for maintenance can be roughly divided into three areas. The first area measures productivity with respect to the effective use of resources. These indicators are normally used by planning and include:
· Number of supervisors
· Percentage productivity as measured by work sampling
· Actual cost versus estimated cost
· Actual manhours versus estimated manhours
· Maintenance staff as a percentage of the

5

total number of maintenance employees.

The second group of indicators could be classified as those used to maintain control of the work force. They are normally used by maintenance management and include:
- Number of work orders scheduled to be closed out versus those actually closed out
- Percentage of overtime
- Backlog, either for the entire work force or for specific results
- Maintenance material requests not filled by the storehouse
- Maintenance stock turnover
- Downtime
- Percentage of maintenance work which is included on work orders
- Percentage of maintenance manhours which has been planned
- Percentage of preventive maintenance
- Maintenance material cost versus maintenance labor cost
- Preventive maintenance manhours versus total maintenance manhours.

The third and last group of indicators includes those which are normally used by plant management to monitor the entire system. They include:
- Number of maintenance employees
- Total maintenance cost
- Maintenance costs versus production costs
- Maintenance cost related to a unit of production
- Production losses caused by maintenance
- Asset devaluation versus maintenance costs
- Ratio of maintenance workers to production workers
- Total annual maintenance costs
- Maintenance costs relative to sales.

There is much evidence to show that these indexes are widely used. A.T. Kearney, a consultant in measurement of indirect labor, includes in its list of recommended performance reports: the work backlog; the departmental performance (by comparing the actual versus the estimated performance); and monitoring percentage effectiveness in such areas as tool room performance.

Albert Ramond and Associates, another well known consultant, use, in their automated management system, labor activity and work-order overrun exception reports (among others) in the management control module.

H.B. Maynard, in "Maintenance Management" [2], recommends the use of standards comparison as an indicator of productivity.

Badger, in its Total Maintenance Service package, recommends the use of standards to evaluate the productivity of the preventive maintenance system, as well as the planning and scheduling system.

The data from these indicators by themselves do not inform management. The comparison of specific data with those data recorded in the past will identify trends which would be of great assistance in determining whether the actions taken by management have, in fact, increased productivity.

A special word should be said about the work sampling process. This is a program whereby the entire work force is periodically screened to determine the percentage of the time that each worker is performing those activities for which he has been employed. Many plants continually sample their work force to determine this percentage of activity. Again, similarly to what has been said previously about trends, the raw value of the "percentage productive" category for work sampling is of little use unless compared with some previous reading. The trends toward increased productivity or decreased productivity are more meaningful if compared with a number of previous readings than if those values are taken independently.

WHAT THE INDICATORS TELL US

To interpret any set of indicators satisfactorily, it is necessary to establish a level of maintenance for the plant in its entirety, as well as for each individual cost center of the plant. Cost centers are normally designed around geographic processing units. In addition, cost centers must exist for generic types of equipment, such as pumps, compressors, major instrumentation, and so forth, in order to be able to analyze the downtime data sufficiently so that an efficient preventive maintenance system can be designed. In addition, the definitive cost centers are necessary to prevent one operator from writing an abnormally large number of work orders for relatively low priority work in the anticipation that management wants him to do this. In that particular instance, the level of maintenance of the particular cost center is artificially elevated. If another operator assumes command of the unit and he writes work orders in a more realistic manner, then the differences in the amounts spent for maintenance must be recognized as a difference in the level of maintenance. In order to prevent this, management must establish certain policies to assure that the level of maintenance throughout the plant truly reflects the desires of management. If a production unit is important to the management, then the level of maintenance should rise; conversely, if a unit becomes less necessary, the level of maintenance should decrease somewhat, although this is not a linear relationship.

Some of the suggested indicators would reflect a need for training. When a craftsman performs a work order and a significant number of those work orders become "rework", then this is an indication that the craftsman is not performing the work in an acceptable manner. In many instances this is an indication of a lack of training and must be recognized as such. Similarly, inability of the maintenance administration to plan and schedule the work force properly is an indication of the need for training and planning and scheduling activities.

Work sampling studies frequently indicate that an abnormally large amount of idle time is accumulating. This might be an indication that more supervision is required or that the task assigned to the supervisors should be reallocated to maximize the time that they have to attend to the different work sites. If this does not prove to be a problem, then it is possible that there are not enough supervisors for the many work sites which require administration.

By keeping a record of the number of "stock-outs" (those times when the storehouse is requested to furnish material and that material is not there), an indication of the ability of the storehouse to service the maintenance function can be realized. A continuing assessment of this ability is an indication of how well the storehouse serves the maintenance function and, as such, could serve as a productivity indicator for the storehouse.

The maintenance function should continually be taking steps to increase the "uptime" of the equipment which is under their care. Maintenance must make sure that the time which is attributed to equipment in the down condition is truly maintenance-caused and not operator-caused. Once these categories of downtime have been successfully segregated they may be used to reflect the ability of maintenance to respond to the needs of operations.

CONCLUSION

All maintenance systems need to be periodically monitored and evaluated. This can be accomplished by periodically evaluating productivity indicators and by use of an annual report to inform maintenance management and top management of the results of any activities to improve productivity of the maintenance work force. The indicators mentioned here form the heart of such a report in that they enable management to quantify the improvement that is being made through the various activities designed to increase the productivity of the maintenance work force.

ACKNOWLEDGEMENTS

The author wishes to thank the Technical Publishing Company for their kind permission to reprint Tables 1 and 2 from *Plant Engineering*.

REFERENCES

1 Chemical Week, McGraw-Hill, New York, July 25, 1979, p. 48.
2 "Maintenance Management," pamphlet printed by H.B. Maynard and Co., Inc. 2040 Ardmore Blvd., Pittsburgh, PA, 15221, p. 19.

Copyright © 1980
TAPPI. Reprinted from Proceedings of the Technical Association
of the Pulp and Paper Industry, 1980 Engineering Conference,
Book 3, pages 421-426, with permission

A Better Way to Use Work Sampling to Measure Maintenance Labor Productivity

Thomas L. Arnold
Manager, Industrial Engineering
ITT Rayonier Inc.
1177 Summer Street
Stamford, Connecticut

ABSTRACT

This is a project aimed at finding better ways to measure maintenance labor productivity.
The project is a joint effort of the Maintenance Management and Maintenance System Subcommittee of TAPPI. In this paper, the author describes the need for measurement in an
improvement program and the pros and cons of using work sampling as a measurement tool.
Some suggestions for overcoming the disadvantages of work sampling are made. The mechanics
of setting up and making the study are also covered, as is the use of the data once it is
collected. Finally, the scope and work plan of the project are summarized.

INTRODUCTION--WHAT WE'RE TRYING TO DO

The material below complements the information contained in Max Brower's presentation, located elsewhere in this publication. Max covers the history of this project, objectives and methodology and other phases of the work not contained here.

Summarizing for those of you who haven't had an opportunity to read Max's paper, this project has as its target the development of a better way to measure maintenance labor productivity. It is a joint effort of the Maintenance Management Subcommittee (a part of the Maintenance and Mechanical Engineering Committee) and the Measurement Systems Subcommittee (from the Industrial Engineering Committee).

The project resulted from a general agreement that there did not seem to be available an inexpensive, well-proven and validated measurement system for maintenance productivity.

HOW EFFECTIVE IS WORK SAMPLING FOR MEASURING MAINTENANCE PRODUCTIVITY?

In brief, the answer to this is that, for the near term, work sampling is probably the best tool that we have. However, it has many serious weaknesses, and there are other techniques which should be more effective for the long run.

The Industrial Engineer has many valid measurement techniques at his disposal. Of these, work sampling is the one in widest use in the paper industry. Evaluation of other techniques suggests that some of these should be more effective than work sampling, though. By the most critical test, "continued use over time these other techniques have generally failed. The reasons for their failure are not well-enough understood at this point in time to know what changes in application are necessary. Developing the methods to resolve these application problems will take time, very possibly up to five years.

8

For this reason, it became apparent to our work group that a dual approach would be necessary. First, since work sampling is already in wide use in the industry, we need to develop a better measurement system based on its use. Secondly, for the longer term, we need to develop a better system utilizing other techniques.

Our presentation here concerns our work on the first objective, that of developing a better work-sampling based system.

THE ROLE OF MEASUREMENT IN AN IMPROVEMENT PROGRAM

Improvement programs won't work without measurement because they require a large expenditure of manpower over an extended period. Management has proven time and again that it will not make this investment without convincing proof of the results. For this reason, bench-mark measurement studies should be made before an improvement program is started and this measurement should be continued periodically during the program.

HOW DOES OUR COMMITTEE'S APPROACH TIE IN WITH THE GENERAL APPROACH TOWARD IMPROVING MAINTENANCE?

The dual approach we have adopted (work sampling first, then a better measurement technique) is very consistent with a cost-effective maintenance cost improvement program.

"Hands-on" work time tends to be very low before any improvement program is initiated. So, the first phase of an improvement program is usually to increase "hands-on" time. We measure this through the use of work sampling. However, work sampling doesn't tell us much about what's happening during the "hands-on" time.

So generally, the improvement approach of ----

1)   Get "hands-on" percent up,

2)   Make "hands-on" time more productive

is quite consistent with our basic approach.

WHAT IS WORK SAMPLING ANYWAY

Conceptually, work sampling is very simple. It involves observation of the activity of the study subject taken at random time intervals. If enough of these observations are made, it can be assumed that the results obtained from the observations are the conditions which normally are present.

The purpose of using work sampling is to draw valid conclusions about the make-up or nature of subject being studied. Normally, the objective of the study is to improve or control some aspect of the subject.

The mathematics of sampling are well-proven, and the technique has been used effectively in most industries and on a myriad of different subjects.

ADVANTAGES & DISADVANTAGES OF WORK SAMPLING

There are several basic problems with work sampling as it is used in our industry. These are listed below with discussions elsewhere. In defense of work sampling though, it also has some very strong benefits. These include:

1.   Simple, straight-forward concept, easily understandable by supervisors most workers.

2.   Simplicity of procedure gives a large, potential observer pool. Don't need specialist.

3.   Widely-used throughout the industry.

4.   Nature of the process allows participation by supervisors in the collection of the data. They can be "involved".

5.   Is effective in establishing relative performance levels for a given craft in a given mill.

6.   Relatively inexpensive to use. However, there are some serious problems associated with its use.

MAJOR PROBLEMS WITH THE USE OF WORK SAMPLING

Our review of a number of sampling procedures in use in the industry suggests that if we are to use the sampling tool effectively, much work must be done. Right now, the sampling programs are generally not effective.

The problems we note must be resolved if work sampling is to function as well as it can for all of us. However, there is enough work required so that probably none of us individually will have the necessary time to accomplish it. Working as a group though, we _can_ do it.

Major problems in work sampling-based systems for maintenance include:

1. Variability and inconsistency of results.

2. Sensitivity of the technique to management pressure.

3. Lack of established yardsticks to judge performance by.

4. Lack of analytical procedures leading to improvement results.

5. Failure of the process to produce information which directly points at what needs to be done.

REDUCING VARIABILITY AND INCONSISTENCY IN SAMPLING STUDIES. Reduction of variability and inconsistency can be achieved through several means. _First_, procedures need to be much more complete than most are now. _Secondly_, management must be convinced that pressing for "high" numbers distorts the results. _Third_, the studies should be controlled and audited by someone not a part of maintenance to assure objectivity.

In most sampling programs, the procedures are not detailed enough, especially in regard to certain critical points. Small procedural variations at these critical points produce large variations and inconsistency in the numbers "that come out the end."

These critical conditions need to be identified and procedurally covered in fine detail. _AND_ a conscious effort must be made to assure that the procedures are followed to a letter for these points. The development of such a procedure is an objective of our committee. A first draft has been finished.

One of the questions which comes up at this point is usually "If results are so variable, why don't people complain about them?" The answer very simply is that the studies aren't being used in most places to produce any results. They're just being read and forgotten. Very little is happening as a direct result of them.

DEVELOPING PERFORMANCE LEVEL SAMPLING PROCEDURE YARDSTICKS. One of the problems caused by the variability of the sampling procedure from mill to mill and company to company is that we can't use the numbers for developing performance guidelines.

Individual mill studies tell us only how we're doing now against how we did in the past. The studies don't answer the important question "Where should we be?"

If all mills and companies used the same standardized, controlled procedure, we could develop some general guideline numbers and also identify conditions which seem to be associated with good productivity.

This, also, is an objective of our committee.

DEFINING THE ANALYTICAL PROCEDURES FOR USING STUDY RESULTS. Most of the systems we have reviewed do not have good detailed instructions which lead from the study results toward necessary action, and then on to improved results. The result of this is that there is little direct action generated by maintenance work sampling studies. A general discussion of analysis is included later in this paper.

CORRECTING THE PROBLEM OF STUDY RESULTS NOT POINTING DIRECTLY AT WHAT NEEDS TO BE DONE. Unfortunately, there is some question as to whether this problem can be corrected any way except through better measurement techniques. This is why our subcommittee has the development of systems as its long range objective.

WHAT ACTIVITIES ARE INVOLVED IN THE TAKING OF A WORK SAMPLING STUDY

I would have to talk for days to cover this topic completely. Fortunately for you, we don't have days.

Therefore, I'm going to broadbrush this subject and recommend that, if you want to explore this in greater detail, you read one of the many books on this subject OR, better yet, attend one of the many seminars being given on this.

A particularly useful and practical book which ties together cost improvement and work sampling is by Professor Wallace Richardson of Lehigh University. It is entitled Cost Improvement, Work Sampling and Short Interval Scheduling.

The American Institute of Industrial Engineering has been sponsoring a seminar on Maintenance Management which also covers the subject very well. The Maynard Management Institute in Charlotte, North Carolina, also has excellent courses in work sampling.

Generally speaking, the following steps describe the work sampling process:

1. Assign the responsibility for the study to someone.

2. Determine the specific objectives of your study and how much manpower (and other resources) can be devoted to it.

3. Inform all personnel, both union and management, of the study and its objectives.

4. Design the study:

   - Select observation categories and define them very well.
   - Choose the number of observations and observers needed.
   - Choose the length of the study.
   - Design the forms needed.
   - Plan the observation routes and procedures for determining where personnel should be.
   - Arrange the procedure for the selection of random start times and routes.
   - Set up procedures for determining the number of people working during the day.
   - Select output measures and develop methods of collecting information to determine them.

5. Inform personnel again of the study including any necessary procedural information.

6. Select and train observers

7. Make a "dry" run to test the study arrangements.

8. Start the study, cross check observa-

tions for consistency and keep running tally of data.

9. Calculate final results.

10. Analyze the results, collect additional information as needed and develop conclusions.

11. Make recommendations for improvement actions.

12. Implement the recommendations.

13. Determine the timing and frequency of future studies.

As mentioned earlier, our committee has developed the first draft of a sampling procedure which will ultimately be submitted to TAPPI for acceptance as a TAPPI STANDARD PROCEDURE.

Copies of our draft will be available to any person who requests one. There is a form here at the front on which you can leave your name and address if you would like a copy when it's available.

ANALYSIS WITH WORK SAMPLING STUDY DATA

O.K.! As a result of the Work Sampling Study, you now have a set of numbers. What do you do with them?

You need "Action" information. The basic test of the value of information is "What does it tell you to do."

This same basic approach also applies to the use of the data from the Work Sampling Study. The objective of the analysis is to 1) determine if things are better, worse, or the same 2) identify areas where additional work or information is needed.

Emphasis must be not on blaming people for problems, but on specifically what is causing them, and what can be done to eliminate them.

Beware of reaction to the numbers in a "first study." The "hands-on" percents normally stun management because they look so low. A common reaction is to jump on supervisors and craftsmen. However, low performances are usually caused more by deficiencies in the maintenance management and stores systems than by the workers themselves. It has been estimated that 80% of lost time is caused by poor management, not poor employees.

As a final comment here, it is not a good idea to compare craft-against-craft and mill-against-mill. The validity and fairness of these comparisons is very questionable and decisions based on this can be disastrous.

11

WHAT NOT TO DO WITH THE SAMPLING RESULTS

The most common procedure is for management to look at the numbers and, if the "good ones" are up, pat everyone on the back. If the numbers are down, management raises hell and tells supervision to get back on the ball--quick.

What are the results from this? Essentially, a flurry of activity for a week, and then, back to the old ways. Why? Essentially, because A WORK SAMPLE DOESN'T TELL YOU WHAT'S WRONG OR WHAT NEEDS TO BE DONE. And, to improve a complex operation, you need specific information as to what needs to be done. Usually, the first step should be to collect more information before doing anything.

A second reaction is to shake up the organization by demoting, promoting, or moving various maintenance personnel around. This is a problem for two reasons. 1) The sampling study does not give accurate or complete enough information on which to make these decisions AND 2) Taking action in this manner guarantees that the maintenance staff will do everything to get high numbers (rather than "correct" ones) during the next study.

WHAT SPECIFIC NUMBERS SHOULD YOU LOOK AT?

The number that tells the overall story is "hands-on" work percent. The "bottom line" in maintenance is the time when the craftsmen physically have their "hands-on" the job that is in progress There are other necessary activities such as getting parts and tools, looking at blueprints, walking to the job, etc; but, REMEMBER--"HANDS-ON" TIME IS THE "BOTTOM LINE"; IT'S WHAT WE'RE REALLY PAYING FOR.

Therefore, "HANDS-ON" work is the only productive work we're interested in when we review the results.

There's another reason for ignoring the "other" work categories to a large degree. They are subject to interpretation on the part of the observer. If the pressure is on the observer, he can easily swing these enough to bias the results to the point of making the study valueless.

Examples of this problem are: 1) Interpreting any walking as "walking loaded" (productive) 2) Interpreting any talking with anyone or any writing as productive. Many times these are personal or idle.

By contrast, "hands-on" work is very difficult to misinterpret unless such misinterpretation is done purposely. Therefore, any evaluation should key on "HANDS-ON" work.

WHAT LEVELS OF "HANDS-ON" WORK SHOULD YOU EXPECT TO SEE?

The levels will vary depending on the craft and the effectiveness of your maintenance organization. No levels have been identified as "good" performance yet. This is one of the important objectives of our project.

However, it is not unusual to see an overall "hands-on" level of 25-30% in initial studies before substantial improvement work. Levels above 45% are very unusual and worthy of suspicion.

WHAT SPECIFIC ITEMS SHOULD YOU LOOK AT OTHER THAN "HANDS-ON" WORK

Specific items which should be examined include those categories which are not productive. Major among these are walking (of all types); waiting/idle and personal time.

WALKING TIME. The acceptable amount of walking depends on a number of factors, but can be roughed out for a given operation mill. Normally, if the total for walking loaded and empty is more than 10-15%, it's worthwhile to investigate further. Some typical causes of excessive walking are poor job dispatching, incomplete planning and excessive centralization of maintenance shops. To reduce walking, you must first know the causes.

WAITING, IDLE; PERSONAL TIME. Since this is completely non-productive, the reasons for these losses need to be pinned down accurately. It is not unusual in many studies for these categories to account for 35-50% of total observations.

A note of caution here. People often think that they know the reasons for lost time and do not collect facts about it. Actual studies turn up surprise after surprise about the real causes. No matter how sure you are of the reasons, a study should still be made.

IF A NON-PRODUCTIVE ITEM LOOKS TOO HIGH, WHAT SHOULD YOU DO?

Since the sample only gives a general distribution of personnel activities, you are left with a problem. The problem is that, although you now may know from the sample that a given craft was idle 20% of the time, you don't know why. The next step is to FIND OUT WHY!!

How to do this? Generally speaking, you are going to have to grub out the answer through a lot of digging and hard work. Usually, special studies must be set up and run. This is where most users of work sampling fall down. THEY DON'T FOLLOW UP the study with the MANPOWER TO GET RESULTS. It takes money ($$$) to make money ($$$). And it is astounding how much time it takes to really find out what's happening and to figure out what should be done about it. It's usually a good investment, though.

One way to dig out more information is to do the sampling over again; however, this time restrict it to only the areas and crafts where you want more information. Whenever observation of a non-productive activity is made, the observer can stop and ask the craftsman what the cause is. The observer then records the given reason.

It is amazing how few rounds have to be made to identify the major causative patterns.

WHEN YOUR SAMPLE NUMBERS HAVE IMPROVED BY A LARGE AMOUNT OVER THE FIRST STUDY, AND THEY'RE STABILIZING, DOES THIS MEAN YOU'RE AS GOOD AS CAN BE EXPECTED?

Remember, the sample gives you only the degree to which your troops are involved in hands-on work. The sample does not tell you:

- whether the work really has to be done.
- the correctness of the method being used.
- whether the crewing is appropriate.
- whether the craftsman skill is good.
- whether the work pace is adequate.

Summarizing the above! Even though the hands-on percent becomes much higher, you may still be wasting $$$ if you:

- have to fix things which should have been corrected or shouldn't have happened at all.
- have craftsmen who use the wrong tools and methods to fix something.
- send too many people to do a job (most common abuse -- send two to do a one-man job).
- have craftsmen whose skills are inferior, AND/OR
- have people who work at too slow a pace for the type of work being done.

So, in a sense, work sampling can only get you to the point where you have managed to get your people's hands' on work more of the time. Resolving any of the above problems requires different programs and different sets of actions.

OUR MAINTENANCE WORK SAMPLING SURVEY

As a starting point to find out exactly what the industry is doing relating to work sampling, we have designed and distributed a questionnaire to a number of mills. A copy of this is attached.

We hope that if your company or mill hasn't received one of these questionnaires, that you will take this opportunity to respond. All respondents will receive a copy of the questionnaire results.

Hopefully, by the time of this meeting, enough responses will be available to present preliminary conclusions.

SUMMARIZING OUR WORK PLAN INCLUDES THE FOLLOWING

1. Industry Practice Questionnaire. Get as many responses from our questionnaire as possible, summarize and distribute the results.

2. Standardized Procedure. Guided by the questionnaire response, develop a practical standardized procedure which can be used by any mill.

Test this procedure in volunteer mills and finalize. Have it approved as a TAPPI STANDARD procedure.

Distribute to as many mills as possible and request that they use it.

3. Performance Yardsticks. From the numbers produced through use of the standard procedure, develop yardstick performance levels by craft.

4. Conditions Favoring High Productivity. Identify conditions underlying good productivity and develop data to support the effects and $$$ value of these conditions.

5. A Better Measurement System. In the long run, develop a measurement system which eliminates the shortcomings of the work sampling-based system.

Help us on this work------and you'll help yourself.

REFERENCE

1. Richardson, Wallace J., Cost Improvement, Work Sampling, and Short Interval Scheduling, 1976, Reston Publishing Co., Inc., Reston, Va.

CHAPTER 2

PREVENTIVE, PREDICTIVE AND ON-CONDITION MAINTENANCE

Reprinted courtesy of Warren Lee, Oceanside California

Organization and Management of a Preventive Maintenance Program

By Warren Lee

A. INTRODUCTION

The common failure in initiating a new PM program is to assume that the systems work and procedures are the essence of the program. Actually most PM activities can proceed with little support from systems work or procedures. The primary need in a successful preventive maintenance program is for problem solving and analytical ability. No PM system or procedure will appreciably reduce maintenance costs or equipment down time unless it is accompanied by problem solving ability.

In other words, most PM systems and procedures merely help identify opportunities for improvement and measure improved performance. The actual improvement can only result from avoiding or solving problems. Problems can be avoided or solved only by people with suitable training, ability, and experience, applying themselves to the problem. In fact many of the desirable results of a PM program can be obtained merely by assigning suitable people to solve obvious problems.

However effective such a group may prove at solving or avoiding problems certain PM activities such as "inspection" and "fixed interval Repair Scheduling" can only be economically established on the basis of reliable records. To provide such records and to help measure the effectiveness of the entire PM operation, certain systems and procedures are desirable. For full effectiveness these systems and procedures should be integrated into a total maintenance system that provides planning and control for all maintenance functions.

Before trying to develop the PM systems and procedures suitable for a specific situation, a new PM group would find it prudent to determine just what might be accomplished immediately by direct attack on known or easily determined problems. Developing the organization and assigning suitable people to this function has several advantages:

1. It usually produces the results (sometimes dramatic results) needed to enlist and maintain management support of a program that otherwise may be slow in providing measurable results.

2. The problem oriented people will be available to act on the problems identified later by the PM system and procedures.

3. The more obvious problems have a good chance of yielding the greatest returns.

The rest of this review is concerned with (1) Defining Preventive Maintenance, (2) Detailing each activity that should be considered in a PM program, (3) Outlining the elements of suitable PM systems and procedures, and (4) Relating PM systems and procedures to those required for effectiveness of the entire maintenance operation.

B. DEFINITION

Preventive maintenance has been defined as any activity that:

    REDUCES EQUIPMENT FAILURES
    REDUCES THE MAGNITUDE OF EQUIPMENT REPAIRS OR REPAIR COSTS
    REDUCES PRODUCT LOSS OR PRODUCTION DOWNTIME DUE TO EQUIPMENT
       FAILURE OR REPAIR
    REDUCES DETERIORATION IN THE PRODUCTIVE CAPACITY OF EQUIPMENT

This definition of preventive maintenance needs one constraint. Any preventive maintenance activity must be economically justified. It must either save money by reducing the long-term cost of equipment or make money by reducing costly interruptions to productivity.

Preventive maintenance activities are of seven general types. The procedures and systems for preventive maintenance are more easily examined if we consider each type of maintenance activity individually.

AVOIDING PROBLEMS

    AVOIDING PROBLEMS BY DESIGN AND MATERIAL CONTROL
    AVOIDING PROBLEMS RESULTING FROM OPERATING PROCEDURES
    AVOIDING PROBLEMS BY PROPER LUBRICATION
    AVOIDING PROBLEMS THROUGH INSPECTION TO IDENTIFY INCIPIENT FAILURE
    AVOIDING PROBLEMS BY SUBSTITUTING SCHEDULED MAINTENANCE FOR
       BREAKDOWN MAINTENANCE

SOLVING PROBLEMS

    SOLVING PROBLEMS IDENTIFIED BY ANALYSIS OF RECORDS
    SOLVING PROBLEMS IDENTIFIED BY ANALYZING EQUIPMENT OR
       REPAIR OPERATIONS

While it is common for Managements to think of preventive maintenance as comprising, simply, either inspection or the substitution of scheduled maintenance for breakdown maintenance, the other listed activities may yield equal or greater savings.

"Solving problems identified by analyzing equipment or repair operations" is here arbitrarily included with other PM activities although the subject contemplates design change or equipment replacement. Some may prefer to think of this as "corrective" maintenance.

Effective preventive maintenance systems and procedures frequently are different for each of the above activities. For each activity, the problems and the work to be done will be discussed as a basis for reviewing systems and procedures that have proven effective.

C. AVOIDING PROBLEMS BY DESIGN & MATERIAL CONTROL

The prevention of excessive maintenance costs begins with the design. Equipment should not be selected on the basis of first cost alone. Consideration should be given to potential operating and maintenance costs. Maintenance costs will be affected by such things as:

Equipment standardization
Parts interchangeability
Service accessibility
Selection of equipment and materials known to perform well in the local environment.

Everyone agrees that these areas of equipment costs deserve consideration. However, few organizations actually consider their effect on costs. Why?

It has been our experience that companies fail to consider the effect of maintenance on equipment purchase decisions because no knowledgable individual is assigned the responsibility for insuring this consideration. Assigning the responsibility is the first step. Developing knowledgable estimates of the cost is the second. The elements of cost that should be considered in each of the above areas will be listed here for guidance.

EQUIPMENT STANDARDIZATION simplifies the mechanic's job. Every equipment line has its own idiosyncrasies. The fewer lines of equipment you ask a mechanic to learn, the more expert he is likely to become. The training period and training effort will be reduced and the work force will have greater flexibility for job assignment. The requirements for supervision will be less. Few organizations do an adequate job of providing the maintenance mechanic with the details of the manufacturer's repair instructions. With a limited variety of equipment, the mechanics eventually learn these details. With a large variety, mistakes are more frequent. More time is lost looking for information.

Although it is difficult to put a dollar value on the maintenance cost savings that result from a craftsman's intimate familiarity with equipment, they can be substantial. Good mechanical work always means longer intervals between repairs.

PARTS INTERCHANGEABILITY that results from equipment standardization also reduces maintenance costs. It reduces the number of spare machines or spare parts that must be purchased and warehoused. For instance, standardizing on a pump or valve, means a single spare part or unit may be stocked for emergency. Alternatively an installed unit in a low-priority application may be borrowed to replace failed equipment in a critical area.

Savings resulting from reducing the complement of required spares usually amount to about one fourth of the purchase cost of the spares annually. Over the course of years, failure to standardize on designs at every opportunity can result in substantial hidden increases in maintenance costs that appear as rising stores costs. Accounting oriented managements attacking this symptom frequently take actions that eventually result in loss of production due to lack of spare parts availability.

MATERIAL CONTROL can also affect maintenance costs. In spite of major advances in specifying the proper materials for difficult processes, considerable art is still involved. Conditions from plant to plant usually vary enough to require changes in some of the materials used for process equipment. Purchases of new equipment should reflect the knowledge of materials gained by those responsible for maintenance.

SERVICE ACCESSIBILITY is another area of design control that is primarily a preventive maintenance function. Most designers responsible for equipment layout in a plant know better than to locate an exchanger where interference will prevent pulling a tube bundle. Many know to provide access for cranes to handle the bundle. But how many are familiar enough with the heavy equipment used by plant maintenance to provide for the maintenance organization the type of access that will be needed when all equipment is in simultaneous use during a major shutdown? How many are equipped to analyze the comparative cost to maintenance of renting extra pieces of equipment during shutdowns versus the cost of built-in lifts and trolleys? How many are familiar enough with maintenance procedures to have a clear idea of the space needed for accessibility during a big shutdown?

Service accessibility isn't only the space needed to pull a tube bundle. It's space for the oiler to get in to check or oil a remote lube point. Without that space, management may have difficulty getting the point lubricated. Accessibility is lube points or inspection points that the operator or service man can get to without removing a guard or safety cover. Poor accessibility not only can increase direct maintenance costs, it can increase downtime and the inefficient maintenance labor associated with emergencies.

To some it may seem that these elements of design and material control are really engineering functions. And maybe they are. However, adequate control is seldom found in plants where the maintenance function has little influence over design. Design and material control is an important area for preventive maintenance.

SYSTEMS AND PROCEDURES for establishing effective preventive maintenance influence over designs and materials are simple. They require the following elements:

1. Management must agree that the concept of maintenance influence over design decisions is desirable.

2. The maintenance organization must provide personnel of suitable ability, training and experience with adequate time to review the designs. This commonly means an engineer experienced in maintenance engineering.

3. The responsible engineer must study the effects that failure to standardize has on those costs discussed here. Where possible he must put an estimated dollar value on those costs. Where this is not feasible, he must marshal his facts with such coherancy that they receive due consideration.

D. AVOIDING PROBLEMS RESULTING FROM IMPROPER OPERATING PROCEDURES

Operating procedures are a fertile field for the prevention of excessive maintenance costs. Most maintenance departments are ready to recite numerous examples of major maintenance expense resulting from improper operating or start up procedures.

Start-up failures, whether on new processes or existing equipment, are common. As an example, it's an unusual petroleum or petrochem plant that hasn't exposed one or more heat exchanger bundles, during start-up or operation, to a high temperature product without the other side of the tube being flooded with the cooling medium. The tubes expand and warp. Troubles begin. The bundle is much more difficult to clean. It doesn't get cleaned as well. Therefore, production may be reduced. The bundle may have to be pulled and cleaned more often. It's a problem until it's re-tubed. Is this a problem that should be left to operations? Seldom is an operating organization able to eliminate this problem on its own. It's a problem in preventive maintenance.

How frequently do rebuilt pumps or compressors run less than a day? What's the cause? It's not always an operating problem. Sometimes maintenances alignments and clearances are wrong. But many problems of this sort can be traced back to improper operations. What should be done with a problem like this? Who should be responsible for solving it? Isn't this a problem for preventive maintenance?

Inadequate operating procedures rightfully are problems for operations management. They must develop the necessary feeling of proprietorship in their employees. They must train and supervise them.

However, the preventive maintenance function also has a responsibility. It can identify the problem areas, alert management to their existence and help develop solutions. This is a delicate undertaking. It must be approached with understanding, not emotion. While it is absolutely necessary to uncover the root cause of the problem, it must be done in a way that allows those who must change their ways to understand the necessity for the change. They can't embrace a change if they feel the purpose of any part of an investigation was to harass them or to fix blame. The investigator must thread his way through all the facts dispassionately and objectively and come up with a procedural approach that operations can recognize as valid.

Eventually such a program should yield written standard operating instructions for start up, continuous operation, shut-down, lubrication, emergencies, and testing. Almost inevitably more instrumentation will be found to be desirable. This may take the form of indicating, recording, controlling or automation equipment.

Accomplishing this function requires manpower and the assignment of responsibility. The responsible organization should be the preventive maintenance group. The individual performing this function must be endowed with exceptional tact and considerable mechanical analytical ability.

The substantial reduction in required maintenance that can result from improved operating procedures dictates that this be a concern of the preventive maintenance program.

E. AVOIDING PROBLEMS BY PROPER LUBRICATION

There probably is no field where one may prevent excessive maintenance costs as easily as in the field of lubrication. The results of a program for improving lubrication are seldom immediate and dramatic. They may even be indiscernable if past records are poor. This is unfortunate, because new programs for preventive maintenance frequently need a few dramatic accomplishments. Nevertheless, this is an almost certain source of pay dirt.

Oil company surveys have indicated that possible 50% of industrial equipment breakdowns can be traced to improper lubricants or misapplication of the lube program.

A consultant's extensive analysis of electric motor failures has shown that one third of them are due to lubrication and bearing failures. Probably there is no more straight-forward lubrication problem than electric motor bearings. There are few of the difficulties associated with lubricating, say, a compressor. Yet we find many failures. Primarily these are due to inadequate control over the lubrication function.

Compressor lubrication presents more complex problems. There are problems of oxidation and pollution that are more difficult to identify and control. Oil on the cylinder walls oxidizes from contact with hot air. On water-cooled compressors, temperature rise due to scale formation increases the oxidation. Oxidation eventually forms insoluble gums that become carboniferous. These deposits usually collect on discharge valves preventing seating. Hot air leaking back raises valve and cylinder temperatures, further increasing oxidation rates.

Solving lubrication problems on a compressor takes more than a regular oil change. It takes technical knowledge of the processes that are taking place and analysis of all observed factors.

The opportunities for preventing excessive maintenance costs range in complexity from lubrication programs for electric motors to those for turbines and compressors.

Other problems preventing good lubrication are more mechanical in nature. Grease ports may be inaccessible. Machinery may be damaged and unable to retain the lubricant. An inordinate number of lubricants may make delivering the proper lubricant to the job difficult and time consuming or may result in application errors. Inadequate lubricating equipment may make the lubrication man's job difficult. The absence of mechanical lubricators in demanding locations may increase the work load uneconomically.

An effective lubrication program requires the following instructions:

    ASSIGNMENT OF SUITABLE PEOPLE
    ANALYSIS OF REQUIREMENTS
    EXPLICIT INSTRUCTIONS FOR EACH JOB
    A REPORTING SYSTEM
    LUBRICANT CONTROL AND STORAGE
    ELIMINATION OF PROBLEMS THAT PREVENT GOOD LUBRICATION

1. ASSIGNMENT OF SUITABLE PEOPLE is critical to an effective lubrication program. In larger plants this means an engineer to analyze the problems. It also requires suitable people to do the work. Lubrication may be handled by maintenance people or by operators. Lubrication by operators has the advantage of maintaining an operator's sense of proprietorship over his equipment. Also, because operations staffing levels are usually determined by emergency requirements, operators commonly have available time, so the lubrication gets done at minimum cost while the operator develops more familiarity with his equipment. The disadvantage is that some operators may not consider it their work and may do it poorly.

   Assigning lubrication to a maintenance man has the advantage of tight control. He has no other primary responsibility to divert him from lubrication. He can become very expert and at the same time monitor the equipment closely for incipient problems. Because he reports to the same organization as the lubrication engineer, there seldom is a problem in getting the work done properly.

2. ANALYSIS OF REQUIREMENTS for a lubrication program includes reviewing and establishing with Production those items that are to be covered. The equipment to be lubricated must be selected. Lubrication points should be identified with stickers that specify the type of lubricant and the frequency.

   Problem equipment and problem locations should be identified and the problems eliminated where possible. Inaccessible locations or locations under safety shields should be brought out or automatic oils should be applied. Temperature problems should be alleviated and damaged equipment that won't hold oil should be repaired.

3. EXPLICIT WRITTEN INSTRUCTIONS should be developed for each lubrication job. These should cover frequencies, quantities, types of lubricant, equipment used and routes followed.

   The lubrication engineer should instruct the operator or maintenance man in exactly what he is to do so that questionable situations will be uncovered early. The procedure should be reviewed again after the lubrication man has an opportunity to standardize on the work.

4. LUBRICATION CONTROL AND STORAGE is an important element of a lubrication program. The goal is to deliver to the bearing the proper lubricant in uncontaminated condition. Four functions are of concern:

   a. SELECTING THE PROPER LUBRICANT
   b. PROTECTING LUBRICANTS FROM CONTAMINATION
   c. DELIVERING LUBRICANTS TO THE POINT OF USE
   d. LUBRICATION PROCEDURES AND REPORTING

   a. SELECTING THE PROPER LUBRICANTS requires a considerable knowledge of both lubricants and lubricating problems. The goal is to stock as few kinds of lubricant as possible to minimize in-plant problems of getting the proper lubricant to the point of use. The general approach is to specify premium quality oils. Not only do they last longer and provide more protection to the equipment, but a premium oil by virtue of its better characteristics is able to replace a wider range of oils.

Buying premium oils may sound like a way to spend money rather than save it. However, many countervailing factors are at work here. Lubrication costs consist of both labor and oil. The cost to apply lubricants is commonly about five times their purchase cost. Premium oil lasts longer and requires less frequent oiling. It also reduces the labor for applying because fewer types must be handled as a result of the premium oil serving a wider range of applications. Limiting the number of oils also provides savings from volume purchases and reduced warehouse stocks.

Selecting the proper types of lubricant to "stock" is a decision for an expert. Frequently it's helpful to call in your supplier for a survey. This can usually reduce the number of commonly used oils to less than a dozen with two or three extra oils for specialized services such as extreme temperature, high contact pressures, corrosion protection, etc.

When the lubricants have been selected they should be identified by the purpose for which they are to be used rather than by the manufacturer's designation. Then the lubrication locations should be identified with stickers that identify both the type of lubricant and the application interval. These stickers are usually obtainable from your dealer.

b. PROTECTING LUBRICANTS FROM CONTAMINATION is frequently overlooked. Organizations select good oils and establish good lubrication schedules and assume the oils as applied will be the same as when they were manufactured. Improper storage of oils can result in contamination by dust, water or inadvertent mixing of different types of lubricants. Lubricants should be controlled from purchase to use. Containers should always be well-marked, covered and stored under roof. Providing special storage facilities helps convey the importance of preventing contamination.

c. DELIVERING LUBRICANTS TO THE POINT OF USE is not as simple as it sounds. Point of use is here defined as the bearing or wear surface. You are concerned with getting the lubricants from the warehouse to the point of use. This may include breaking down bulk shipments into area supplies, delivering the lubricant to the machine that needs it and possibly providing an automatic oiler to store it and feed it to the bearings.

Two principles are of concern. The delivery system should accomplish the desired result of getting proper lubrication to the wear surfaces, and the system should be economical. This is an area that frequently requires novel thinking. Sometimes the most difficult problem can be solved with a very simple approach. Consider the following example of a novel approach:

Rock drills in a mine operate under very severe conditions. Over the years designs have been developed that stand up and drill well so long as they are properly lubricated. Past practice has been to carry 5-gallon cans of oil into the mine and tell the miner to oil the drill regularly. It was hard to get the oil close to the heavy drill. Miners were not particularly concerned about caring for equipment. As a result, the drills didn't get oiled and the failure rate was high. The solution was to package drill oil in pocket sized plastic flasks that the driller could carry into the mine each morning.

Your purpose in delivering oil to the point of use is to make it easy for the man responsible for lubrication to do his job. There are many ways to accomplish this besides packaging the oil in a pocket-sized flask. The lubrication man should be furnished with elaborate carts with dispensing pumps and a scavenger tank and pump if these will help him do his job. Controlled sub-stores should be provided and resupplied as needed to prevent lubrication personnel establishing their own stocks that too often take the form of an open grease pot or an uncovered oil can. The details of getting lubricants delivered should be worked out with the individual responsible for doing the work.

As soon as a lubrication program is working satisfactorily, it is time to review each of the applications to determine where it is economically feasible to install centralized, automatic oiling systems. These come in many styles and types. They include mechanical oilers that operate with the machine, clock-controlled oilers, etc. Also valuable are alarm devices and interlocks that alarm oil failures or prevent start-ups under inadequate lubrication. Such retrofit installations may be justified where the cost of a lubrication failure is high in damage or lost production or where there is a requirement for the frequent application of oil.

d. LUBRICATION PROCEDURES AND REPORTING need to be well-conceived. Procedures should be documented and tell the oiler precisely what he should do and what he should not do. For functions the oiler should perform daily, a permanent instruction should be written. For oiling to be done at weekly or longer intervals, the instruction should be issued each time the work must be done. Both types of instruction should provide a checkoff for the oiler to indicate that he has completed the work. The best way to handle this usually is by work order.

The oiler should be trained to report unusual situations found during his lubrication duties. These should be reviewed by people capable of evaluating the problems and directing a suitable course of action.

One man, preferable a lubrication engineer, should be responsible for the program. He should develop lubrication schedules, lubricant specifications, and the written procedures based on his analysis of the requirements for lubrication. He should check the work occasionally to see that it is being performed properly. At intervals, he should review the procedures to determine if they can be improved or simplified.

F. AVOIDING PROBLEMS THROUGH INSPECTION

We have discussed three types of preventive maintenance activities that
fall under the general classification of avoiding problems. The fourth is
inspection. This is a subject on which feelings run high. Some people
believe all inspection is wasteful, that the need for repairs can be statistically forecast and the repairs made without wasting money on inspection. At the other end of the scale is the late L.C. Morrow, formerly
Chief Editor of "Factory". He devoted the chapter on Preventive Maintenance in his Maintenance Engineering Handbook almost exclusively to
Inspection. What inspection functions can reasonably be expected to yield
good results?

Failures of mechanical equipment are seldom instantaneous. Usually there
is a period of operation when the equipment gives one or more signs of
incipient failure. If maintenance is alert enough to read these signs,
they often can act to reduce the cost of breakdowns.

1. INSPECTION has as its purpose the discovery and correction of unfavor-
    able situations in the developing stage in order to prevent breakdowns.
    Reducing breakdowns may increase production as a result of greater
    equipment availability. Maintenance costs may be reduced because
    repairs can be made before major damage is done. Overtime may
    decrease, and labor problems may be simplified because repairs can be
    scheduled during normal working hours. The pay-off for inspection
    comes from corrective action taken before failure occurs.

    What are the usable signals of impending failures? Here are the six
    primary sources:

        a. VIBRATION
        b. NOISE LEVELS
        c. OIL CONDITION
        d. VISIBLE LEAKS
        e. AREA CLEANLINESS
        f. OPERATING VARIABLES such as Temperature, Pressure and Flow

    Consider what inspection can do in each of these areas:

    a. VIBRATION ANALYSIS AND INTERPRETATION is a fruitful way to initiate
        an inspection program. Probably few periodic inspection routines
        have the potential for substantial savings demonstrated by vibra-
        tion analysis.

        Because effective vibration analysis equipment is relatively new,
        experience with it is limited and its value is not commonly recog-
        nized. However, some users have achieved results that can only be
        described as fantastic.

        For example, one large maintenance department with excellent equip-
        ment records could compare costs on electric motors that were
        allowed to run to failure, that were serviced on a time interval
        basis with replacement of bearings and that were serviced when
        vibration analysis indicated it was needed. Major savings resulted
        from monitoring electric motors alone.

When vibration readings were taken and plotted, they could establish the time at which a failure could be expected within days (sometimes within hours). This allowed them to get maximum life from an installation and still schedule maintenance. They also avoided catastrophic failures.

The development and implementation of a vibration monitoring program has several elements:

Suitable equipment is necessary.

An adequate staff must be established and trained.

The equipment to be monitored must be selected, and vibration reading points must be established.

Initial reading intervals must be determined and monitored until experience establishes the time curves with which equipment problems are reflected by changes in vibration.

A record-keeping, plotting, interpreting and reporting system must be developed.

Analytical responsibility must be assigned.

b. NOISE LEVELS are the second source of information for inspectors. This type of signal may be monitored by vibration equipment, with a stethoscope or by ear. Most people have heard a dry bearing squeal, oiled it and seen it go on to operate for a long life. But how many have used stethoscopes to hear the first whisper of a leak in compressor valves so that maintenance could change only bad valves during a scheduled shutdown?

Vibration manifested as noise offers an easily-detected warning of trouble. When Operations learns maintenance has a man to listen to unusual noises with a stethoscope, he will get plenty of calls. Out of this will come early warning of some incipient failures.

Some organizations believe you should never assume the only way to analyze noise is with a vibration analyzer. They believe in sending an expert with a listening device. Operators understand that approach. Soon operators become a crew of full time inspectors that work for nothing and alert the expert as early as a warning is available.

The primary requirements for a good noise monitoring program are a good and inquisitive mechanic backed up with a vibration analyzer and an organization that will take action when difficulties are uncovered.

c. OIL CONDITION may be the oldest and most ignored indicator of incipient trouble. Almost any problem with a bearing or lubricated surface manifests itself as a change in oil condition. A few inspection procedures have been developed around this concept and are being used today.

A refinery on the west coast developed an oil reservoir lubricant condition indicator for pumps. This was a plastic cylinder piped to connect to the oil reservoir. When the oil in the bearing is clear, the oil in the indicator is clear. At the first sign of trouble with the bearing, the oil in the indicator darkens. They found that signs of impending failure were in advance of actual ruined pump shafts and damaged housing fits. On the other hand, failure to heed the warning from the oil condition indicator always resulted in destruction of the bearing.

Bearings were detected that had a chip on the ball or that were running in a cocked attitude so they wouldn't track properly. Bearings were found that had a tight spot. Sleeve bearings were found in the initial stages of wiping. New and repaired equipment that checked out O.K. under routine examination (no excessive noise, vibration or heat) was found to have bearing problems that were only signalled by changes in oil condition.

They believe the signal comes from cracking of the lubricant resulting from a hot spot caused by chipped or skidding balls. They found that in bearings damaged by water or with poor shaft fits, the oil turned brown. They attributed this to abrasive in the housing lapping the bearing to eventual destruction.

This refinery managed to develop all this information from a simple tell-tale attached to oil sumps as a lube oil condition indicator. Consider what an aid the indicator was to inspection. It could be seen a hundred feet away. Every operator automatically became an inspector.

It is a responsibility of Preventive Maintenance to develop effective ways of using every known indicator to anticipate and avoid trouble.

d. VISIBLE LEAKS are another sign of impending trouble. There are all kinds of different practices on leaks. Some plants have area men who, among their other duties, fix leaks as quickly as they appear. In other plants steam leaks blow for days. When maintenance gets around to fixing them, flange faces are wiredrawn or valve stems are cut out to the point that a minor task becomes a major repair.

e. AREA CLEANLINESS is a tip-off to impending trouble. One refinery tank farm inspector concerned with always finding steam trap effluent around the bottom of crude oil storage tanks found that piping and drainage changes reduced excessive repairs to corroded tank bottoms to a small fraction of previous levels.

f. OPERATING VARIABLES are another source of early warning of impending failure. Heat exchanger problems make themselves known this way. Sometimes monitoring can be handled by a computer that can calculate the deterioration in heat transfer conditions on a shift or even hourly basis and thus help pinpoint obscure operating procedures or water treatment practices that cause the difficulty.

Most equipment problems signal their development by temperature increases. This is particularly true of compressors and bearings. It's good practice to feel of equipment regularly to detect temperature changes. However, temperature rise usually signals a late stage of the developing problem.

Preventing excessive maintenance by inspection for the purpose of identifying changes in operating temperatures, pressures and flows is generally an operations responsibility. However, it is a way to reduce both the cost of maintenance and the cost of production losses resulting from equipment downtime. If maintenance department studies indicate the department is not receiving warning of impending trouble soon enough to schedule repairs while they are still minor; if all Operation's requests for equipment repair are emergencies because of major equipment damage; then it's time for Maintenance to gather the information that will convince Management that closer inspection by Operations is needed.

2. INSPECTION OF EQUIPMENT is a function over which there is honest disagreement in many plants. The loudest complaints come from plants where the inspection procedures require the equipment to be shutdown for partial disassembly and thus cause loss of production. The complaints arise because maintenance has failed to prove their inspection program is economically sound.

While Maintenance can frequently prove that preventive maintenance inspections requiring the disassembly of equipment result in lower over-all equipment costs, the big dollars usually are connected with maximizing production. Where maintenance costs are small compared to the value of product produced, it may be better to run equipment to breakdown or even destruction rather than interfere with production.

The point is that Maintenance must consider the economic effect on production of any preventive maintenance or inspection program. No program that requires interrupting production should be initiated until its potential economic soundness is proven.

As an alternative, every effort should be made to find effective on-stream inspection procedures. For example, Radiographic equipment usually can be used on stream. The new water cooled transducers for ultrasonic equipment permit on-stream inspection that can define layers of scale, coke and sludge as well as wall thickness of pipes and vessels.

The computer can be used in some cases to optimize inspection intervals. For example on pressure vessels the computer can feed current readings into ASME Code equations to calculate minimum thickness, maximum allowable pressure, corrosion allowance plus corrosion rate and percent. On the basis of these calculations, the computer can determine whether the shell or either head is controlling, and establish the next inspection date based on half life.

The first step in developing an inspection program that requires shutdown of continuous process equipment is to gather records that measure the extent and cost of downtime due to equipment failure. The next step is to evaluate all approaches that might reduce these failures short of scheduled shutdowns for inspection and repairs. Only when this approach does not yield adequate on-time should scheduled shutdowns be considered.

A program of periodic inspection must be based on economic factors. The inspection procedures must be justified by reduced downtime, reduced maintenance costs, or both. The elements of an effective program must include:

Establishing the equipment items that are to be inspected and fixing responsibility for performing the inspection.

Determining the type, extent and interval of inspections required for the selected equipment. An engineering analysis is desirable to assure that the relationship between inspection costs and probably profit improvement is optimized.

Existing check lists or practices should be reviewed with production and maintenance. The procedures should be optimized and standardized before they are published.

Establishing inspection intervals and procedures should be controlled through a work order and a job scheduling system. Inspections should be initiated by and reported on maintenance work orders. A schedule should initiate the release of the work order.

Continual review and optimization of inspection intervals and procedures.

It is easy to handle the scheduling and reporting procedure manually with a multicopy "standing work order" and a manually posted "schedule" to show when the work order is to be issued, has been issued and has been returned. Copies of the schedule can then serve as a management report to show the status of completion of scheduled inspections.

There are advantages to putting the scheduling and reporting system on a computer. While this will provide little cost savings over a manual system, various reporting permutations and combinations became available, including exception reporting.

G. SUBSTITUTING SCHEDULED MAINTENANCE FOR BREAKDOWN MAINTENANCE

Scheduled maintenance is another preventive maintenance procedure that can reduce maintenance costs. It is nothing more than programmed adjustment, repair or replacement of parts. Why bother to inspect something ten times and finally repair it if such maintenance can be programmed on a statistical basis to provide better results or lower cost than inspection?

The key concept is economics.

Probably the most widely embraced scheduled maintenance procedure is the practice of changing fluorescent bulbs on schedule. The life of these bulbs is consistent enough that it is economical to change out the entire complement of bulbs on a scheduled basis. This proves cheaper than replacing individual bulbs when they fail.

One organization, finding electric motor failures expensive, calculated that they could replace electric motor bearings on a two-year interval (one year on some difficult applications) and be money ahead because they would avoid damage to shafts and windings. An excellent approach.

After they got under way with their program, they found that the savings weren't what they had expected. In fact, motor repair was costing more under the new program than under the old. Further probing indicated that some number of the repair motors were failing catastrophically after running for a few days or weeks. Because of shop problems which they were unable to control at that time, they went back to servicing motors only when necessary. Fortunately vibration-monitoring equipment became available about that time. With vibration-monitoring they were able to get maximum service out of a motor between repair jobs while catching most problems early enough to prevent catastrophic failures. The savings from this approach proved greater than those calculated for the original approach.

Scheduled maintenance has another advantage secondary only to avoiding shutdowns. Scheduled work permits getting needed materials, tools and equipment to the job site and coordinating the mechanics so the work can be done more efficiently. Thus, scheduled maintenance is always <u>less costly</u> than the same work done on an emergency basis. It may prove less costly even when extra jobs must be undertaken to shorten the intervals between scheduled repairs enough to insure against costly breakdowns.

H. <u>CORRECTIVE MAINTENANCE - SOLVING PROBLEMS IDENTIFIED IN THE FIELD</u>

CORRECTIVE MAINTENANCE is a critical function in any effective PM program. Specifically, it is the analysis of repetitive or expensive repair jobs to find ways to eliminate them or to reduce their cost.

The corrective action required commonly involves design modifications, but it may only involve changes in maintenance or operating procedures.

There are two types of activity involved in corrective maintenance: 1) selecting the problem areas to work on and 2) developing economic solutions to the problems. The method of selecting the problem areas to work on deserves careful consideration because some rather elegant systems have failed to produce tangible results.

Maintenance problems are selected for analysis in one of two ways. (1) By directly observing equipment in operation or under repair and considering what might be done to reduce its susceptibility to failure or to reduce the costs for repairing it. (2) By analyzing equipment repair or downtime records for leads to major cost items which then can be analyzed for improved approaches.

It is this latter method (analyzing equipment repair records to identify problem areas) that has come under occasional criticism. Usually when criticism has developed, it has been because the program was started with a non existent or inadequate record base. As a result, there was a lapse of time before the records became effective in identifying problems. When a maintenance program is slow about producing results, the maintenance organization is in danger of losing management support.

There is no excuse for a PM operation to get into this difficulty. If records are not immediately available there are other ways to get leads on repetitive or expensive repair jobs. Almost any mechanic or engineer that has been around for a period knows the equipment that is critical from an operational or repair point of view.

The crux of any corrective maintenance operation is to get solutions to existing problems. In setting up any sort of new PM program, it is desirable to parallel any other functions with an engineer who immediately goes into the field to look at equipment or repair work that is known to be expensive or repetitive. The five or ten major concentrations of maintenance expense are usually known. A good man may score some dramatic successes. This not only eases the way for the total program, but, handled correctly, it can generate enthusiasm among the craftsmen.

Actually, there are few maintenance problems for which some craftsman hasn't figured out some solution. A good engineer can clean up these solutions, see that the craftsman gets full credit from management for the idea and cause a turn-about in the plant work force so that everyone strives for improvement.

A few examples of successful problem-solving may provide guidance for your effort in this area:

A mine had considerable trouble with air drill breakdowns. They had extensive records showing maintenance cost as a function of footage drilled. A good engineer read the drill manufacturer's service manual and discovered they were operating at too low an air pressure. The manufacturer indicated that this would limit the impact on the drill and thus reduce the footage drilled but it would not reduce the wear and tear on the drill. He increased the air pressure, brought about an immediate increase in footage drilled and in footage per repair dollar, and became an instant hero.

A petrochem plant replaced expensive alloy plug valves like pencils. Corrosion seemed uncontrollable until the PM engineer found a special silicone lubricant.

Another petrochem company spent thousands of dollars yearly replacing acid-eaten cement floors before their PM Engineer found a satisfactory Epoxy flooring material.

A smelter replaced the flexible coupling on a hammer mill almost monthly until their PM Engineer proved a V-belt drive would absorb all that shock.

A refinery reduced boiler water treating costs substantially with a new electrostatic scale control system.

A contractor eliminated thousands of dollars of annual repair and downtime cost for Bulldozer clutch repair by bringing the grease fittings out so they could be greased by a man standing beside the machine rather than sliding on his back in the mud underneath the dripping engine pan.

A refinery substantially reduced the number of contractor people hired to unbolt flanges and exchangers during shutdowns by providing impact wrenches for every other man.

An engineer eliminated thousands of dollars of annual maintenance clean-up of coked-up vessels by proving that the unit could produce more product on the continuous operation made possible by a lower feed rate that did not coke-up the equipment.

A turbine with a history of substantial cost of repairs resulting from start-up damage ran for years without a start-up associated failure after the installation of an instrumented, automatic start-up system.

Expensive high voltage power system outages in a smelter environment were almost eliminated by a program for regularly spray cleaning the insulators.

All of these are examples of preventing maintenance costs through the medium of solving existing problems. Leads to most of these problems could be obtained without recourse to equipment records. They result primarily from an inquisitive person asking the question, "How could I do this better?"

If all this can be accomplished without equipment records, why are they needed?

I. CORRECTIVE MAINTENANCE - SOLVING PROBLEMS IDENTIFIED BY ANALYZING EQUIPMENT RECORDS

Good equipment records are necessary not only to help identify certain problem areas but also to provide a measure for performance.

Equipment records may be kept manually or by computer. It is usually possible to develop adequate information from either approach. The manual system consists primarily of an individual file for each important item of equipment. Into this file goes each work order or service report for the equipment. The size of the file identifies the most troublesome items of equipment. The material in the file provides the background for an analysis of the problems.

A much better equipment record and reporting system can be developed around a computer. The computer not only can collect and arrange information in the most efficient form for analysis but it can run correlation studies between maintenance costs and other variables of interest to a PM program such as the cost center using the equipment, the craftsman repairing the equipment, failure reasons, cost of failures, etc.

There are several effective approaches to identifying equipment problems by means of records. None are valuable unless you are organized and staffed to solve such problems as are identified.

J. DATA ACCUMULATION AND REPORTING

Preventive Maintenance requires a good system of records and reports. P.M. usually must be sold to Management. And rightfully so. It's possible to embrace program elements that aren't economically sound. Shutdowns or repairs on a cyclical basis may result in more lost production than break-down maintenance. Uncontrolled lubrication or inspection programs can be costly. P.M. records serve three functions:

> They measure the performance of the PM function.
> They identify opportunities for improvement.
> They help avoid uneconomic programs.

Depending on the size of the plant, the need for sophisticated performance measurement and the data processing equipment available, the data system may be manual or computer oriented. It is usually possible to develop suitable information from either approach.

THE INPUT DOCUMENT is basic to any information system. Most P.M. programs use the existing maintenance work order for information input. Coding is usually provided to identify both the class of equipment and the individual equipment item. In addition, Failure Reason may be coded together with downtime or the cost of lost production.

In manual systems, the work order may be posted with all labor and material charges. In computerized systems material requisitions may be charged against the work order in conjunction with cost accounting and material stock control procedures.

IF A COMPUTER is available, various reports may be obtained from the "file" information very simply through a report generator. These can consist of summary reports of the monthly or annual maintenance expenditures on each item of equipment or an exception report listing only those items on which maintenance expense exceeds a certain percent of initial cost or a certain dollar level. The computer can run correllation studies between maintenance cost and other variables such as the cost center using the equipment, the craftsman repairing the equipment, failure reasons, cost of failures, etc. Quarterly or annually the computer can provide a one time summary listing of all work orders accumulated against each item of equipment.

IF NO COMPUTER is available, the completed work orders may be filed by equipment number to furnish a complete history for review as necessary.

PROBLEM EQUIPMENT can be identified most easily by computer manipulation and listing of file data on an equipment or class of equipment basis. However, manually filing each work order by equipment number also will usually identify problem equipments. The more work orders that go into a file, the thicker the file gets. This is the so-called "Fat File" system of problem identification.

Once problem equipments have been identified and the magnitude of costs has been ascertained in terms of maintenance or lost production, maintenance management will know where to apply their P.M. efforts.

AN EQUIPMENT NUMBERING SYSTEM is fundamental to the equipment maintenance costs records necessary for good control of P.M. The equipment number should be in two parts: A two-digit "class of equipment" code and a four-digit "item" number.

Location is considered by some as a usable bit of information to code. However, if used, it should be coded separately on the work order rather than being combined with the equipment code. It is desirable to be able to track a piece of equipment wherever it is moved within the plant.

The equipment numbering system should be suitable to the needs of operations, accounting, fixed asset records and engineering in addition to the administration of the P.M. program. Equipment brought under the system should be permanently "tagged" with the correct equipment number.

A NUMBERED FILE FOLDER should be provided for each numbered item of equipment. Into each folder should go all information relevant to the piece of equipment, such as Equipment Data Sheets, spare parts lists, manufacturer's recommendations, check lists, cost information, correspodence, etc. Manual information systems should include signed-off copies of completed work order. This file becomes the source of specific advice and technical information for those charged with analyzing the P.M. program.

Using this information and studies of the equipment itself, they will be able to:

1. Identify recurring or high cost repair situations. Search for fundamental causes, and for the most economic alternative solutions.

2. Adjust the frequency of Preventive Maintenance work so optimum equipment availability is attained with least annual cost of ownership.

3. Review the job content (manhours, materials and tools) of completed Preventive Maintenance Work Orders to refine the applicable "Work Description" in order that only the needful job content for each level of maintenance work is specified.

SCHEDULING OF PM work orders should be handled in the same manner as scheduling for other maintenance work. Daily scheduling is most effective. The work orders specifying the work to be done by each crew should be issued the preceeding day. All material procurement equipment schedules and coordination between crafts should be planned for.

The scheduling can be done clerically or by computer. It is not difficult for a clerical group to introduce PM work orders into the day's scheduling operation. PM work orders should be written no more often than once a year. They should be reproduced and filed in quantity so that a single copy can be dated and released each time the job is to be done.

The release of PM work orders can be controlled by a schedule. The schedule should list each work order on a separate line by title and number and show the intervals at which the job should be repeated. Following this listing, the schedule should provide squares for fifty two weeks where the status of the PM jobs can be shown.

On jobs that must be repeated weekly, a diagonal should be drawn in the square at the time the W/O copy is pulled from the file and released. When the work order is returned marked "completed" the diagonal should be crossed to make an "X" in the square for that week. The schedule form can then be Xeroxed monthly to provide a report on PM performance.

On Jobs that are to be repeated at intervals greater than a week, the next scheduled date can be indicated on the schedule. When that week arrives, the work order can be issued and the schedule marked to reflect that issue. When the work has been completed, the schedule can be marked and the next schedule date entered. This prevents useless shortening of the interval if for some reason a job can't be completed as scheduled. The increased interval will alert management to the presence of some slippage in the schedule.

K. INITIATING THE PROGRAM

Initiating a Preventive Maintenance program requires planning and effort. It must interface with Maintenance, Operations, Engineering and Accounting in many areas. Guidelines must be clearly established. The organization responsible for the program must be suitably staffed. Relationships to other organization must be defined. Organizational authorities and responsibilities must be logically distributed if the group is to be free of inconsequential tasks and able to accomplish its mission.

The PM program can be initiated best if there is an existing planning function that plans and schedules all work. This usually insures an information system that will provide a suitable measurement of PM performance. If planning and a management information system are not available it may be better to undertake the development of a system for Planning, MIS and PM as a single project. It is questionable whether the development of an information system for PM alone is an economical approach.

Usually it is desirable to review both the organization and function of the entire maintenance operation at the time a major P.M. effort is initiated. This permits integrating all maintenance functions to make optimum use of available facilities and personnel.

Many cost-saving programs are probably initiated "in house". However, because of the many efficiencies that derive from effectively integrating all the elements of larger maintenance organizations, an outside consultant frequently can both save the company time and provide a greater profit return by advising on organizational structures and procedures. This is particularly true when the consultant has extensive experience over the entire field of maintenance.

L. STAFFING FOR PREVENTIVE MAINTENANCE

Developing and installing a comprehensive Preventive Maintenance program requires competent and adequate staffing. The staff should report to the Manager of Maintenance and should be technically oriented. Establishing a Maintenance Engineering organization usually proves most effective.

A separate Maintenance Engineering organization's goals, responsibilities, organization and operating procedures should be clearly defined. Performance Measuring systems for the group and for the functions for which they are responsible should be developed. Accounting for the group should be under overhead rather than on an individual project basis.

The responsibility for developing and initiating the program should rest with Maintenance Engineering. They should study the equipment, formulate the program, and set up workable schedules for cyclical inspection and service which can be given to the maintenance scheduling group. The Maintenance forces would then perform the required inspections, adjustments, lubrications and repairs.

M. MEASURING PERFORMANCE

The Maintenance Engineering group should submit periodic reports concerning the efficiency of the Preventive Maintenance Program covering such subjects as:

1. Overall schedule compliance with the general plans of the Preventive Maintenance program.

2. Alternate solutions to reduce the high costs associated with certain units of equipment.

3. Recommended economic studies for equipment retirement, modification, updating, etc.

4. Accomplishments of the P.M. program including:

    a. The effect of Preventive Maintenance expenditures on the total maintenance cost of selected items of equipment.

    b. The effect of Preventive Maintenance expenditures on the downtime of individual equipment items.

    c. The effect of Preventive Maintenance expenditures on the amount of emergency and high priority work on selected equipment.

5. A year-end report covering all aspects of the Preventive Maintenance Program and outlining a specific, detailed program to improve its function and to further reduce overall maintenance costs.

    The group responsible for the Preventive Maintenance Program must always be required to justify their existence on a profit-improvement basis. This requires that economic analysis be a part of every job. When economic justification becomes difficult, the staff should be reduced or returned to engineering jobs where they will be under project accounting rather than in the overhead account.

N. MAINTENANCE MORALE AND MOTIVATION

The most successful Preventive Maintenance groups are those that make maximum use of the knowledge and ideas of other operating and maintenance people. No matter how much experience or inventiveness is possessed by Preventive Maintenance personnel, they seldom can equal the storehouse of information and ideas held by the people who operate or repair the equipment daily.

Preventive Maintenance people can substantially expand their effectiveness by motivating all personnel to contribute ideas for reducing the quantity of maintenance required or for improving the performance of equipment. This requires perceptive listening, identification of every idea's positive features, delicate handling of impractical ideas and recognition for the people that contribute money-saving ideas.

It is a good idea for a member of management to casually thank and compliment individuals for their good ideas. Properly handled, this procedure can snowball until every employee becomes cost conscious and works to help management reduce costs and increase profit.

This may be the ultimate contribution of an effective Preventive Maintenance Program.

Reprinted by permission of the Council of the Institution of Mechanical
Engineers from Proceedings, 1979 Volume 193, No. 13

MANUFACTURE AND MANAGEMENT GROUP

MAINTENANCE MANAGEMENT
AND FAILURE STATISTICS

A. KELLY, CEng, MIMechE
M. J. HARRIS
Simon Engineering Laboratories, University of Manchester

1 INTRODUCTION

Many of the problem situations in maintenance and reliability involve probabilistic variables. The modelling of such situations therefore requires a basic understanding of failure statistics, which we shall here take to be the application of statistical techniques to the description and analysis of the patterns of failure of plant or of its separate components. Such techniques have been used extensively by the reliability engineer (1) and over the last few years a number of papers (2, 3, 4) have shown how failure statistics can directly help the maintenance manager.

In this paper we shall try to clarify the use and limitations of failure statistics in the field of maintenance.

2 THE DEFINITION OF FAILURE

One way of regarding a large and complex industrial plant is as a hierarchy of parts ranked according to their function and replaceability (see Fig. 1). Every component, or assembly of components, has to perform a function. When this performance, by some predetermined criterion, becomes inadequate, the component, or assembly, is said to have failed.

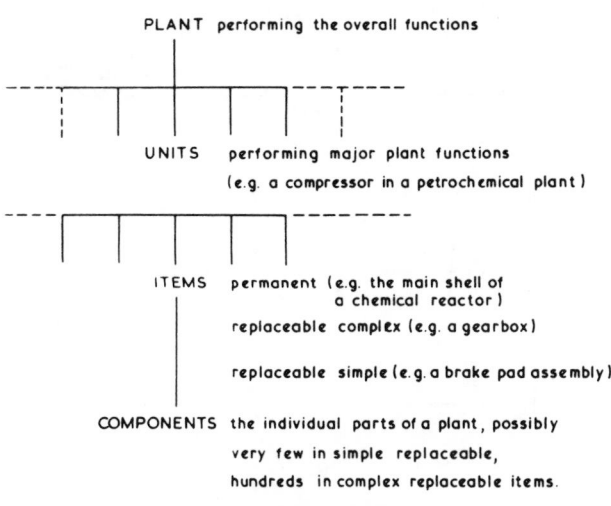

PLANT performing the overall functions

UNITS performing major plant functions

(e.g. a compressor in a petrochemical plant)

ITEMS permanent (e.g. the main shell of
a chemical reactor)

replaceable complex (e.g. a gearbox)

replaceable simple (e.g. a brake pad assembly)

COMPONENTS the individual parts of a plant, possibly
very few in simple replaceable,
hundreds in complex replaceable items.

Fig. 1. Plant hierarchy

3 BASIC FAILURE STATISTICS

Many failure-causing mechanisms give rise to measured distributions of times-to-failure which approximate quite closely to probability density distributions of definite mathematical form, known as *probability density functions*, or pdfs. Such functions therefore provide mathematical models of failure patterns, which may be used in performance forecasting calculations.

3.1 Age dependent failure

A simplified model of an age dependent failure process is shown in Fig. 2a. There is an increasing likelihood of failure as the component grows older and, therefore, the times to failure of a large number of such components would be distributed as shown in Fig. 2b (where $f(t)$ = probability of failing per unit time, at running time t). Such information can be presented in other forms that may be more useful for modelling the failure situation. Figure 2c gives the fraction of items *surviving* at running time t, i.e. the survival probability $P(t)$. Figure 2d gives the age-specific failure rate $Z(t)$. This last is defined as the fraction, $Z(t)$, of those items which have survived up to time t which are expected to fail in the next unit of time.

A failure pattern of this type indicates that the primary failure is age-related, and due to mechanisms such as abrasion, corrosion or fatigue. Often, it approximates quite closely to the well known 'normal pdf' although analysts have shown that for certain specific mechanisms other functions, while broadly similar, may be more appropriate (e.g. for aircraft structural fatigue, the Birnbaum–Saunders family of expressions (5)).

3.2 Purely random failure (negative–exponential pdf)

It is the experience with a very wide range of components and items that, under their normal operating conditions and during their normal operating life, they *do not* reach a point of wear-out failure at some likely time that could be called 'old-age'. On the contrary, a given item is as likely to fail in a given week shortly after installation as in a given week many months later. In short, *the probability of failure is constant and independent of running time*; the item is always effectively 'as good as new'. Very often, such behaviour indicates that the cause of failure is external to the item. A fuse is always as good as new until a short circuit elsewhere in the system blows it. Whitaker (2) quotes a case in the chemical process industry where age-dependent failure of pump seals was caused by gas locking and heat checking due to inappropriate design of other parts of the process flow path.

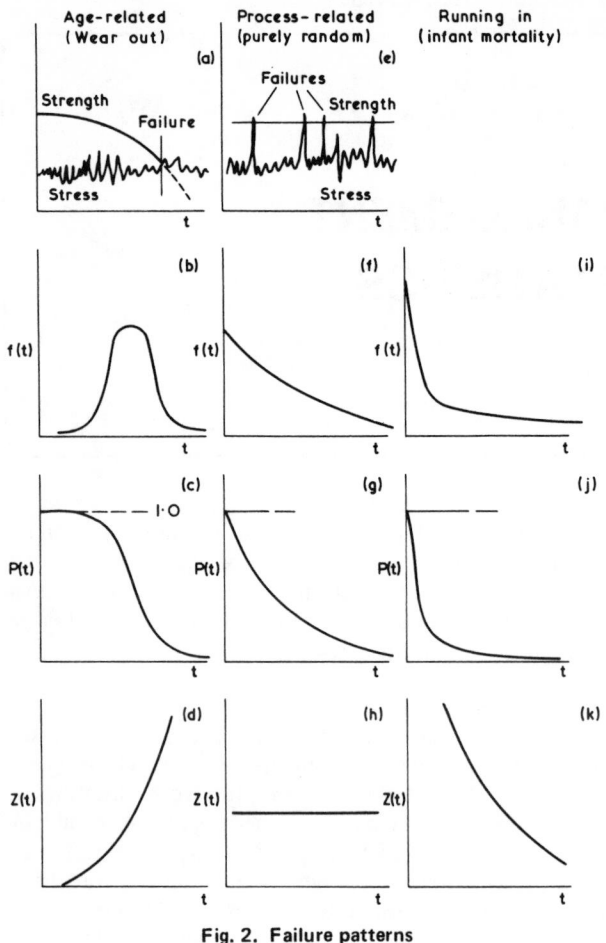

Fig. 2. Failure patterns

A simplified model of this situation is shown in Fig. 2e, the corresponding probability density distribution $f(t)$ of times to failure in Fig. 2f, the survival probability $P(t)$ in Fig. 2g, and the age-specific failure rate $Z(t)$ in Fig. 2h.

A failure pattern of this type indicates that the primary failure mechanism is process-related, e.g. maloperation and/or poor design.

3.3 Running-in failure (hyper–exponential pdf)

With many types of equipment the probability of failure is found to be much higher during the period immediately following installation or maintenance than during its subsequent useful life. Such behaviour results in a pdf of times to failure which, by contrast with the negative exponential pdf which shows a single exponential fall off, exhibits two phases – an initial rapid exponential fall and a later slower exponential fall. This is illustrated in Fig. 2i, 2j and 2k. It is evidence that some of the items are manufactured or installed with built-in defects which show up during the running-in stages. Those that survive this stage without failure were without such defects to begin with; they go on to exhibit the sort of time-dependent failure probability previously discussed. (The equipment is not improving with age! Some items merely start off with a better chance of survival than others).

A failure pattern of this type indicates that the primary failure mechanism is manufacture or assembly-related.

4 THE WEIBULL PDF

Although the conventional statistical techniques discussed above can be used to describe failure patterns the Weibull

pdf $(6, 8)$ is a particularly useful technique because it provides:

(a) A single pdf which can be made to represent any of the three types of pdf's of times-to-failure described in Section 3.
(b) Meaningful parameters of the failure pattern, such as the probable minimum time to failure.
(c) Simple graphical techniques for its practical application.

For this distribution

$$f(t) = \frac{\beta(t-t_0)^{\beta-1}}{\eta^\beta} \exp\left\{ -\left(\frac{t-t_0}{\eta}\right)^\beta \right\},$$

and

$$P(t) = \exp\left\{ -\left(\frac{t-t_0}{\eta}\right)^\beta \right\}$$

Each of the terms in the above expressions has a practical meaning and significance.

The threshold time-to-failure, or guaranteed life, t_0

In many cases of wear-out the first failures do not appear until some significant running time t_0 has elapsed. Age specific failure rate $Z(t)$ is non-zero and rising only after t_0, so in the Weibull expressions the time factor always occurs as the time interval $(t - t_0)$.

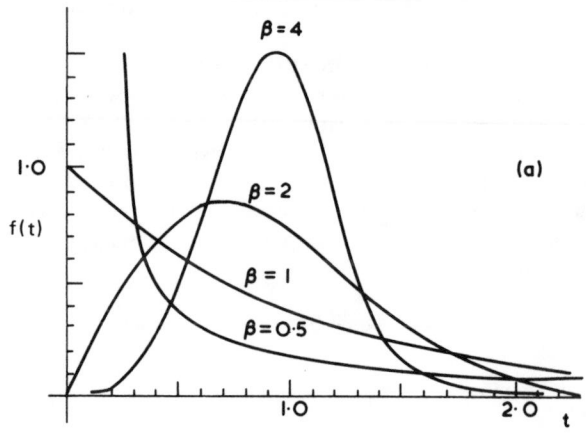

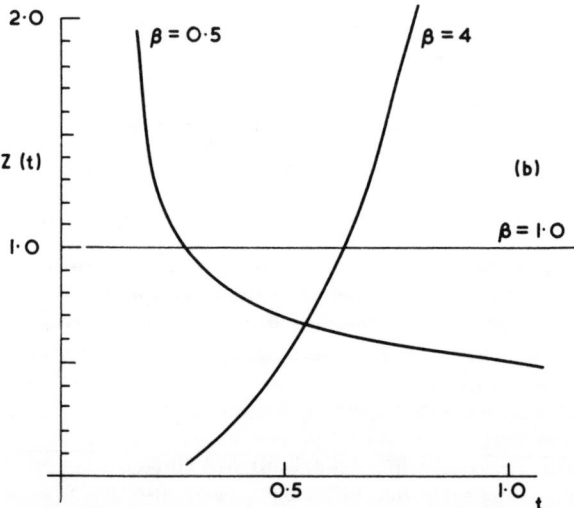

Fig. 3. Influence of shape factor β on Weibull distribution (for simplicity, $t_0 = 0$ and $\eta = 1$)

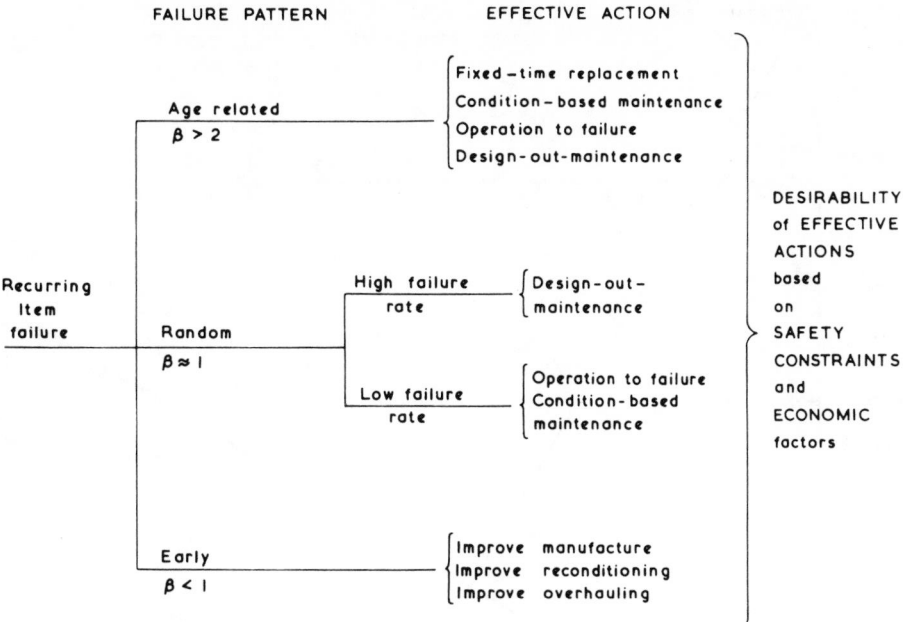

Fig. 4. Decision tree for remedial action, recurring failure

The characteristic life, η

When $(t - t_0) = \eta$, $P(t) = e^{-1} = 0.37$, i.e. η is the interval between t_0 and the time at which it can be expected that 63 per cent of the items will have failed and 37 per cent survived.

The shape factor, β

Figures 3a and 3b show how the various patterns of time-to-failure and of age-specific failure rate are characterized by the value of β. A 'running-in' or 'infant-mortality' failure process is characterized by a value significantly less than one, a purely random process by a value fairly close to one, wear-out by larger values, although if β is less than, say, three then a purely random factor is still significant.

A relatively simple graphical technique, using a specially designed probability graph paper, for the practical application of the Weibull function, is described in more detail in the Appendix.

5 APPLICABILITY TO MAINTENANCE MANAGEMENT

The most important application of failure statistics lies in providing information to designers and reliability engineers, enabling them to estimate system reliabilities, availabilities, expected lives etc., with greater certainty. Failure statistics can, however, also be of use to the maintenance manager directly, and in two main ways, firstly in the *diagnosis* of the nature of a recurrent equipment failure, (see Fig. 4), secondly in the *prescription* of solutions to maintenance problems.

5.1 Diagnosis

An example of this was obtained by the authors during an investigation into the low availability of diesel loaders in a large ore mine. Maintenance management were unsure of the cause of recurrent failures of universal couplings; poor reconditioning was suspected. Analysis of the data (Fig. 5) showed that the Weibull β parameter was close to 1.0, i.e. failures occurred quite randomly with respect to running time so that the possibility of a reconditioning-induced (running-in) failure mechanism (or a wear-out mechanism) could be eliminated. Finally, the cause was found to be

poor design of coupling bolts, which worked loose quite randomly with respect to time. Similar analysis of the failures of associated gearboxes gave a β of about 0.5, suggesting, in their case, a reconditioning-induced problem.

In order to put this use of failure statistics into perspective it must be emphasized that failures in the manufacturing and process industry

(a) can occur for a multitude of reasons (6) (see Fig. 6),

(b) are usually random with respect to time (2), and

(c) can be determined in the vast majority of cases without the aid of failure statistics.

5.2 Prescriptive application

Figure 4 shows that there are a number of effective actions which can be taken when failure pattern is age related. The problem is usually to decide which of these actions is best and clearly there are many factors which influence the decision, and a variety of procedures (e.g. working through a decision tree) which can be brought to bear. A useful procedure in some situations is to use statistical failure data and/or analytical functions of the kind discussed in Sections 3 and 4 to construct costing models to estimate the most economic replacement schedule. This, however, is applicable only to simple items (as previously defined) which exhibit markedly age related failure. Complex items *do not* in general show age related failure and, even if this were so, cost factors usually indicate actions other than fixed time scheduling e.g. condition-based maintenance or operation to failure.

6 THE COLLECTION OF FAILURE DATA

Learning the principles and techniques, of statistical failure analysis, reliability design, or replacement theory, is one thing, collecting and making readily available a sufficient body of dependable data for these purposes is quite another. It cannot be done without adopting painstaking, tightly controlled, and expensive procedures. Broadly speaking, the difficulties involved are of two sorts, arising from human factors on the one hand and equipment factors on the other.

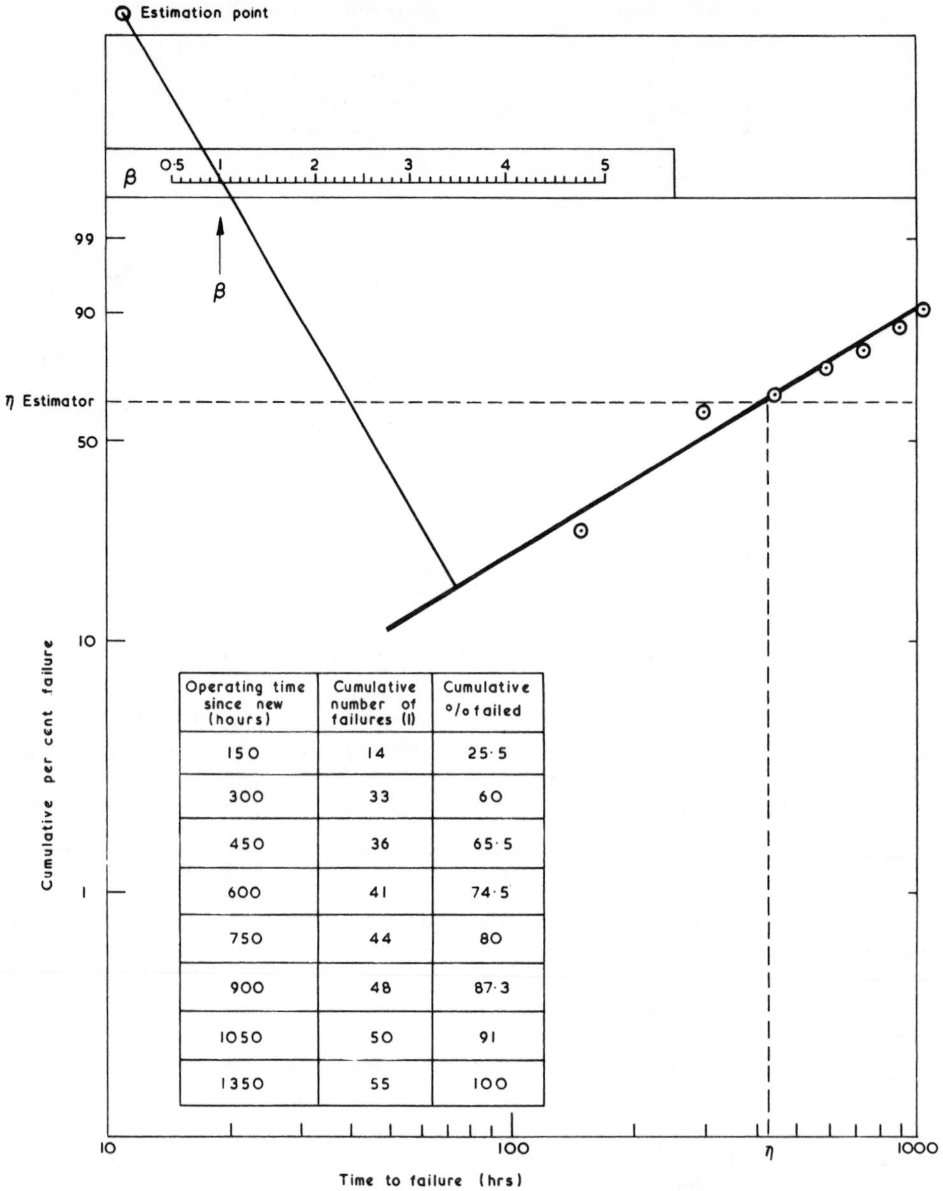

Estimation point

β 0·5 1 2 3 4 5

β

99

90

η Estimator

50

10

Cumulative per cent failure

Operating time since new (hours)	Cumulative number of failures (I)	Cumulative %failed
150	14	25·5
300	33	60
450	36	65·5
600	41	74·5
750	44	80
900	48	87·3
1050	50	91
1350	55	100

1

10 100 η 1000

Time to failure (hrs)

Fig. 5. Weibull plot of times-to-failure; universal couplings

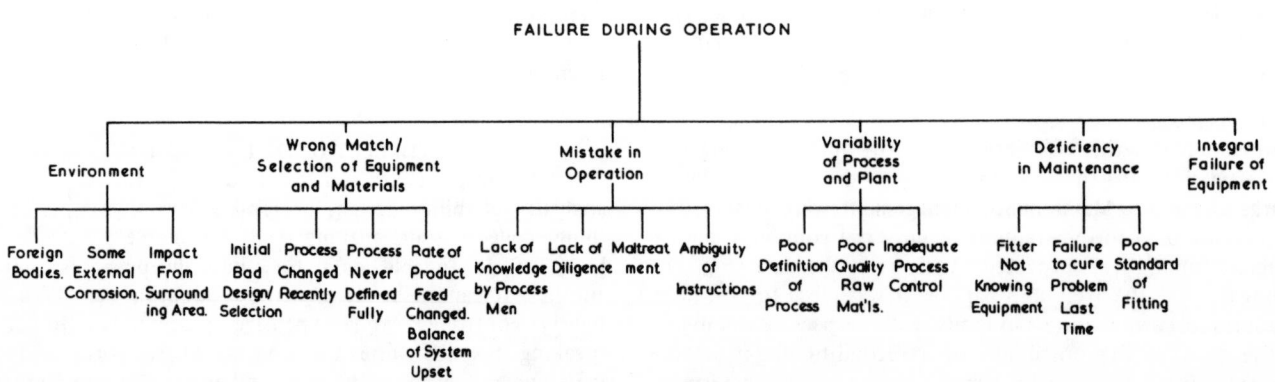

FAILURE DURING OPERATION

Environment — Wrong Match/ Selection of Equipment and Materials — Mistake in Operation — Variability of Process and Plant — Deficiency in Maintenance — Integral Failure of Equipment

Foreign Bodies. | Some External Corrosion. | Impact From Surrounding Area.

Initial Bad Design/ Selection | Process Changed Recently | Process Never Defined Fully | Rate of Product Feed Changed. Balance of System Upset

Lack of Knowledge by Process Men | Lack of Diligence | Maltreatment | Ambiguity of Instructions

Poor Definition of Process | Poor Quality Raw Mat'ls. | Inadequate Process Control

Fitter Not Knowing Equipment | Failure to cure Problem Last Time | Poor Standard of Fitting

Fig. 6. Cause of failures on a batch chemical plant

With regard to human factors it is clear that great resistance exists at all levels (especially at shop floor level, expressed via its representative trade unions) to the collection of failure data. Many data collection systems studied by the authors (6) have been less than successful because of shop floor resistance, or because of apathy and lack of commitment by the immediate management. Often this was no surprise because the system had been badly designed, and pushed on to the maintenance department without adequate prior consultation. Some common faults were as follows:

(i) Insufficient consideration of the motives for data collection, much of which was therefore unnecessary and manifestly so.

(ii) Insufficient appreciation of the problems of analysing the data so as to provide decision-making information to the right people at the right time.

(iii) Over-elaboration and excessive demands on the data collectors.

When designing data collection systems it is vital to aim at the maximum possible simplicity. A system providing limited, but correct, information at the right time is infinitely preferable to one which is sophisticated but unreliable and installed merely for managerial window dressing.

Equipment factors can also prove a serious obstacle. Clearly, data can be accumulated more rapidly, and its subsequent analysis be more fruitful, in those industries (transport, the armed services, communications, forging etc.) in which many identical items, in comparable environments, can be observed concurrently. If only a single item exists, or is available for observation, times-to-failure have to be observed between consecutive replacements or repairs and by the time sufficient data is accumulated its analysis may be of diminished value. Very often the cause of recurring failures can be diagnosed directly and designed out, thus terminating the original study.

7 CONCLUSION

Failure statistics can be used to assist maintenance management in a limited way both to diagnose the cause of failure and to prescribe solutions. The limitations are due to the relatively few cases where the techniques can be applied and to the cost of collecting and analysing the failure data.

As far as the maintenance manager is concerned the main purpose of collecting data must be to identify problem areas and to provide back-up information for on-site investigation.

REFERENCES
(1) GREENE, E. and BOURNE, J. Reliability technology, Wiley, 1972.
(2) WHITAKER, G. D. Statistical reliability models for chemical process plant. Instn. Chem. Engrs, Symposium on Design for reliability, April 1973.
(3) JARDINE, A and KIRKHAM, A. Maintenance policy for sugar refinery centrifuges. Proc. Instn. Mech. Engrs., 167, 1973, 53/73.
(4) KELLY, A. A case study of availability. Instn. Mech. Engrs. Conference on Terotechnology, Sept. 1974.
(5) BIRNBAUM, Z. W. and SAUNDERS, S. C. A new family of life distributions. J. Appl. Prob. 1969, 6, 319–327.
(6) KELLY, A and HARRIS, M. J. The management and techniques of industrial maintenance, Butterworths, March 1978.
(7) BROOKS, D. SEDDON, G. and KELLY, A. Total Technology Project, ICI Organics Division, University of Manchester.
(8) JOHNSON, L. G. Theory and technique of variation research, Elsevier, 1964.

APPENDIX
Weibull analysis of a small and incomplete sample

In most practical situations there may only be the opportunity to measure a handful of times of failure. Indeed, the items under examination might be large, expensive and of low failure rate, and as is likely in such a case, only a few might yet have been made. In addition, some of them might still be running, not having reached the failure point (i.e. 'suspended'), or some of the test may have been terminated (i.e. 'censored') prior to failure because, in their case, the test conditions were accidentally altered. In this situation the results of any analysis will necessarily be subject to greater statistical uncertainty, nevertheless, a Weibull analysis may be required on the grounds, for example, that an approximate result at the end of a fortnight may be of more value than a precise one obtained by waiting for another three months A technique using 'median ranks' as demonstrated in the following example, and illustrated in Fig. 7 is then appropriate. The statistical reasoning underlying the method is fairly sophisticated and for a more full explanation, illustrated by practical examples, the reader should consult reference (8).

Ten oscillating springs are being tested to failure. The situation to date is as follows:

Spring number	Cycles to failure	Spring number	Cycles to failure
1	9 100	6	(7 200)*
2	8 000	7	4 500
3	6 300	8	(5 000)**
4	11 100	9	8 400
5	3 300	10	5 200

*Still running
**Test terminated, inadvertent overspeed

A Weibull fit to the data is required. The procedure is as below.
1. The failure points are ranked in ascending order (column 2 of Table 1) and classified as failed, f, or suspended, (or censored), s, (column 3).
2. For the first failed item the 'new increment' is calculated from the formula:

$$\text{New increment} = \frac{N + 1 - (\text{Order number of previous failed item})}{N + 1 - (\text{Number of previous items})}$$

Table 1. Spring failure data

1	2	3	4	5	6
Spring number	Cycles	Class	New increment	Order number	Median rank
5	3 300	f	1	1	0.067
7	4 500	f	1	2	0.163
8	5 000	s	–	–	–
10	5 200	f	1.125	3.125	0.272
3	6 300	f	1.125	4.250	0.380
6	7 200	s	–	–	–
2	8 000	f	1 350	5 600	0.510
9	8 400	f	1 350	6 950	0.639
1	9 100	f	1 350	8 300	0.770
4	11 100	f	1 350	9 650	0.898

43

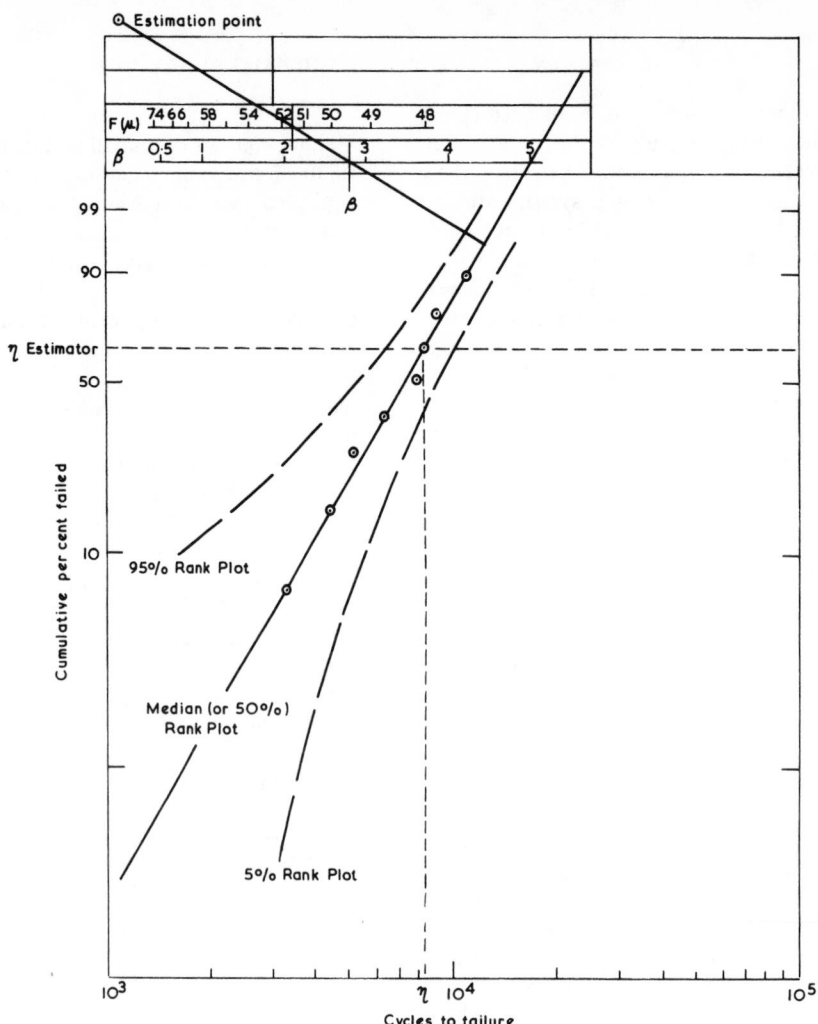

Fig. 7. Weibull plot of spring failure median ranks (data of Table 1; 5 per cent and 95 per cent ranks also shown)

where N = total number of items in the sample (i.e. 10). Since this is the first failed item, the previous order number is zero. Also, in this case, the number of previous items is zero and therefore the calculated new increment is 1 (column 4). Note that if the first failure had been preceded by some suspended items the new increment would have been greater than 1, e.g. if the first two items had been suspended, the new increment would have been $(10 + 1 - 0)/(10 + 1 - 2) = 1.22$.

3. The order number of the first failed item is obtained from the expression:
   order number = new increment + previous order number, i.e. in this case,
   order number = 1 + 0 = 1 (column 5).

4. This procedure is repeated for all the remaining *failed* items, in succession, i.e.

$$\text{second failed item: new increment} = \frac{10 + 1 - 1}{10 + 1 - 1} = 1$$

$$\text{order number} = 1 + 1 = 2$$

$$\text{third failed item: new increment} = \frac{10 + 1 - 2}{10 + 1 - 3} = 1.125$$

$$\text{order number} = 1.125 + 2 = 3.125$$

etc., etc.

The value of the new increment obtained for the first failed item after a suspended item remains constant, and therefore need not be revised, until the next group of suspended items.

5. Having completed column 5, the corresponding median ranks (column 6) are calculated from the formula

$$\text{median rank} = \frac{\text{order number} - 0.3}{N + 0.4}$$

e.g. for the fourth failed item

$$\text{median rank} = \frac{4.250 - 0.3}{10 + 0.4} = \frac{3.950}{10.4} = 0.380$$

6. The median ranks, expressed as percentage, are plotted against cycles (or time) to failure on Weibull graph paper (e.g. Chartwell Graph ref. 6572). The results are shown in Fig. 7 and in this case t_0 is taken as 0 since it gives a good straight line plot. If this was not the case t_0 would be established by trial plots (the straightest plot giving the value of t_0).

7. The characteristic life. η, is the value of t (in this case 8200 cycles) at which the plotted line reaches the 63 per cent failed level.

8. As shown, a line is drawn through the 'estimation point' and perpendicular to the straight line fit. The point at which this intersects the special scale at the top of the

graph gives the value of β (in this case 2.78 approx). Note that this line also intersects another scale which indicates the value of the cumulative per cent failed, F at the point $\bar{t}$. In this case $F = 51.5$ per cent which corresponds to $\bar{t} = 7400$ cycles.

To sum up, the observations fit a Weibull pdf of parameters

$$t_0 = 0$$

$$\eta = 8200 \text{ cycles}$$

and $\beta = 2.78$

We have also deduced that $\bar{t} = 7400$ cycles (the mean life)

9. Johnson **(8)** gives tables of 5 per cent and 95 per cent ranks, for various sample sizes and order numbers. It is a simple matter to plot these, as well as the median (or 50 per cent) ranks, against the observed cycles (or times) to failure, thus obtaining (see Fig. 7) the 90 per cent confidence band for the data, i.e. the band within which it is 90 per cent probable that the plot *obtained from a very large number of items* would lie.

In the absence of suspended items the procedure is simpler, in that the order numbers are simply 1, 2, 3, 4, . . . , N.

This paper is published for written discussion. The MS was received on 28th August 1978 and was accepted for publication on 28th November 1978. 33

Presented at the SME Predictive Maintenance Conference, October 1979

Tuning for Top Performance of the Automated Factory

By Theodore E. Gibbon
Honeywell, Inc.

Tomorrow's CAM system will be extremely reliable. And, if a failure should occur, built-in self-diagnostics will automatically locate and identify the fault, and switch to backup devices, thus permitting uninterrupted control.

Though built-in self-diagnostics will certainly enhance system dependability, it will not be a substitute for an "effective maintenance program"—a program that will not only predict failures before they occur. But, will insure the system is properly tuned and operating at its peak efficiency.

Tuning, why it's important, and the impact it will have on the CAM system will be the topic of this paper.

INTRODUCTION

Tomorrow's CAM system will be extremely reliable. And, if a failure should occur, built-in self-diagnostics will aid maintenance in repairing the fault. Nonetheless, tomorrow's CAM system will still require an "effective maintenance program" performed routinely to assure continuous, uninterrupted, dependable system performance—one which will predict failures before they occur, and which will permit the system to perform the manufacturing process at peak efficiency.

PROCESS CONTROL VS CAM

To predict tomorrow's CAM system maintenance requirements does not require any "crystal ball gazing." Control strategies similar to a CAM system exist today, and have for a number of years, in the Chemical and Petroleum Industries.

The process control systems used today are in many respects analogous with tomorrow's CAM system. As shown in Figures 1A & B, they have similar hierarchies, distributed stand-alone control, supervisory control capabilities and self-diagnostics. Yet, both will continue to require routine maintenance. To discover why, it is necessary first to examine the nature of failures.

THE NATURE OF FAILURES

The components of any system, whether it be a sophisticated process control or CAM system, are subject to three basic types of failures which may occur throughout the system's useful life. These failures, shown graphically in Figure 2, are Early-Life failures, caused typically by component infant mortality; Useful-Life failures, typically caused by random component failures; and Wear-Out failures, which may be accelerated by harsh plant environment and inadequate maintenance.

Many precautions taken in the design, manufacture and testing of today's control systems help minimize these types of failures. These precautions may include the use of conservative design parameters; computer-selected and preconditioned components; automatic test and evaluation of circuit boards in subassemblies; production thermal testing (hot-cold cycling); and final operational system checkouts. There's not much we can do about Wear-Out failures. And, even the best design, manufacturing techniques, test procedures and quality assurance cannot prevent completely Early-Life and Useful-Life failures. Even self-diagnostics cannot completely prevent failures, but self-diagnostics, properly designed, programmed, and implemented can discover, locate, and indicate many potential failures almost before they occur.

There are even some process control systems which, at the first sign of failure, automatically switch to redundant or backup systems, permitting the process to operate uninterrupted or at some level of graceful degradation.

Self-diagnostics, however, cannot detect all changes—changes which may be indicators of impending failures. In fact, there can be a subtle deterioration in overall system performance which self-diagnostics may not detect. Only when the deterioration surfaces as a failure will self-diagnostics detect it, as shown in Figure 3.

Even though this subtle deterioration of system performance which begins the day the system is installed cannot always be detected by self-diagnostics as we know it today, it can be detected by a Predictive Maintenance Program.

Before describing this program, a review of how a system's performance can deteriorate is in order. It is also important to keep in mind that it is not so much the ultimate failure that should concern us but, rather, the impact this subtle deterioration has on process efficiency, production rates, manufacturing costs, and, ultimately, profit.

PERFORMANCE DETERIORATION

The effect of component deterioration on the performance of a CAM system can best be demonstrated by examining a small portion of the system in Figure 1—a portion consisting of an input/output conveyor, lathe, robot for material transfer, and an automatic measuring device. Figure 4 shows a key operation of a CAM system, a valve-plug turning operation. A robot, which off-loads raw stock from an input conveyor, loads the lathe, removes the finished plug from the lathe, and then inserts it into an automatic inspection device which measures the plug. After it is measured, the robot then places the finished valve plug on the output conveyor.

Although each device in the above process stands alone and functions independently from the other, a manufacturing computer supervises and coordinates the entire machining process. This computer, which occupies a higher position in the control hierarchy, is also stand-alone, and it continues to make valve plugs to a specific standard unless it receives instruction from yet another computer controlling both total manufacturing inventory and distribution, as previously shown in Figure 1.

The entire system is programmed so that information is fed from one source to another, and as deviations are evaluated, the proper control action is transmitted to the effective activity as required.

As deterioration occurs, the system does not work as it should. And the results are less output and higher scrap rates.

But enough with generalities. To be more specific, the following are some examples of the many reasons for system deterioration and its probable effect on the previous described valve plug manufacturing operation:

Lack of Cleanliness

- The lack of cleanliness can have disastrous effects. Any blockage of air flow caused by dirty filters or by dust entrapped around an electronic component can cause heat problems and eventually lead to electronics damage. The effects of overheating may remain even though the components affected have cooled. This initial damage can cause failures several months after the original incident. Such failures would certainly result in slowing production. The same damage may also cause a sudden shift in calibration of the automatic measuring device. Such a shift would be misinterpreted by the manufacturing computer as tool wear, causing the computer to make unwarranted changes to the lathe program.

Moisture Retention

- Another cleanliness-related problem is centered on the dirt and oil build-up around components on circuit boards. Not only does this reduce the air flow around the component, but it can retain moisture. Moisture can produce intermittent shorts between the paths on circuit boards and would also result in slowing production.

Voltage Fluctuation

- The power supplies for the lathe, robot control, or automatic measuring devices could develop voltage level changes, ripple and noise. Excessive voltage levels, especially on systems which incorporate integrated circuits, can cause irreparable damage. Low voltage and ripple can cause intermittent failures which slow production. Noise, see Figure 5, in the lathe and robot controls would produce rough cuts and possibly misaligned work pieces.

Frequency Changes

- The clock oscillator controlling feed rates for the lathe and sequencing speed of the robot could change in frequency, resulting in a slow-down of the entire production process. In conjunction with the clock frequency, a change in resolver sine or cosine signal phase relationships could shift, causing position errors both in the lathe and robot.

Feedback Noise

- Feedbacks, whether resolver or photo encoders, could become noisy and change in output level. This noise, as previously shown in Figure 5, is amplified by the servo amplifiers or the SCR's of the motor drive system and would result in erratic machine movement, rough cuts and oscillation of the associated axis both in the lathe and robot.

Servo-Loop Imbalance & Improper Gain

- Servo loop balance and gain adjustments could change, causing both drift and oscillation, as shown in Figure 6, which produces lead/lag errors on contouring systems and over or under-shoot on positioning systems. Both drift and oscillation slow down the manufacturing process and increase scrap rates.

Zero-Switch Maladjustments

- The set point in zero switches could get out of adjustment due to vibration, or accidental movement, causing extensive damage to the lathe and robot drives or the machines themselves.

The accumulated incidences of system deterioration in a complex process control or CAM system hierarchy could be overwhelming. All aspects of raw material flow—from product design to production, from inventory to distribution—could be affected. And, we know for a fact that today's self-diagnostics cannot detect all the indicators of system deterioration and that a Predictive Maintenance Program is a must in order to meet profit objectives.

PREDICTIVE MAINTENANCE AND TUNING

There are basically 3 types of maintenance—Breakdown, Preventive, and Predictive. Breakdown maintenance is performed after a failure occurs. Preventive maintenance includes routine servicing, inspection, cleaning, and adjustments to prevent a failure before it occurs. Predictive maintenance is the periodic monitoring key operating parameters to observe changes which may forecast failures before they occur or deterioration of performance.

The basic premise of a Predictive Maintenance Program is that after a piece of equipment is installed and running, it will acquire its own distinct operational characteristics—its own "fingerprint" as it were. When we record this "fingerprint", we establish a performance standard. We can then periodically measure the performance of the equipment against the standard, and if there is a deviation, determine its cause and take necessary corrective action. This is the fundamental basis of a Predictive Maintenance Program and its associated tuning. In concept, it is like the CAM system control hierarchy or any management system.

THE STEPS INVOLVED

The steps involved in a Predictive Maintenance Program are not complex. However they should be performed by a knowledgeable, capable, qualified technician. There are basically four steps: "Fingerprinting" to establish the standard; "Periodic Measurement" of select parameters against the standard; "Analysis" of any deviation; and "Corrective Action" as required.

Fingerprinting

Key operating parameters are first measured and recorded to establish the device's performance standards. Electronic wave shapes should be drawn on graph paper to show rise times, frequency, spikes, ripple, and other characteristics. Some of the other operating parameters which may be measured and recorded are as follows:

- Power supply voltage levels and the magnitude of any ripple.
- System clock frequency and wave shape.
- Feedback device resolvers, tach, photo encoders signal level and wave shapes.
- System arithmetic computation.
- System input/output timing.
- Axis drive voltage level and wave shape.
- Analog to digital and digital to analog converter accuracy.

Periodic Measurement

On a routine basis, preferably at 90-day intervals, key parameters recorded in Step 1 are measured and recorded again. And any deviation from previous readings are noted. This step helps, not only to identify failures before they occur, but also to detect any subtle deterioration of performance.

Analysis

Any deviation found in Step 2 is analyzed. If the deviation is critical and correctable, its correction provides not only an opportunity to prevent a failure, but to make machine performance improvements.

Corrective Action

The fourth and final step capitalizes on any identified impending failures and machine performance improvement opportunities found in Step 3. It includes the repair and replacement of any marginal part, circuit board or component, and the correction of any other deviations found in Step 3, required to optimize performance, accuracy, repeatability, and efficiency.

The characteristics of such devices as lathes, robots and automatic measuring devices will be in constant flux due to changing component characteristics, normal wear and tear, vibration, and many other factors. Thus, an ongoing Predictive Maintenance Program checking identified variables against established standards will be a must. Otherwise, total system performance may be compromised.

CONCLUSION

In addition to self-diagnostics, a Predictive Maintenance Program which predicts failures before they occur and provides frequent system tuning will be a necessity in tomorrow's CAM system. Such a program will assure continuous dependable service and efficient operation of the manufacturing process. How such a program will be implemented is certainly the option of the users. They may elect to do it themselves or contract it through a reputable maintenance vendor. In either case, it should not be done without the proper system knowledge, test equipment and tools, and by those qualified and trained to do it.

BIBLIOGRAPHY

"Solving the NC Maintenance Problem", A. Webb, MODERN MACHINE SHOP, July, 1979.

"Maintenance, Numerical Control", T. Gibbon, AMERICAN MACHINIST, April 1979

"How Does NC Maintenance Effect Your Productivity and Profit?", T. Gibbon, MACHINE AND TOOL BLUE BOOK, November 1977

"NC Maintenance Contract Reduces Machine Down Time for Valve Manufacture", D.J. Singer, MODERN MACHINE SHOP, May 1977

"Maintenance, Numerical Control", A. Buchanan, NC COMM LINE, January/February 1977

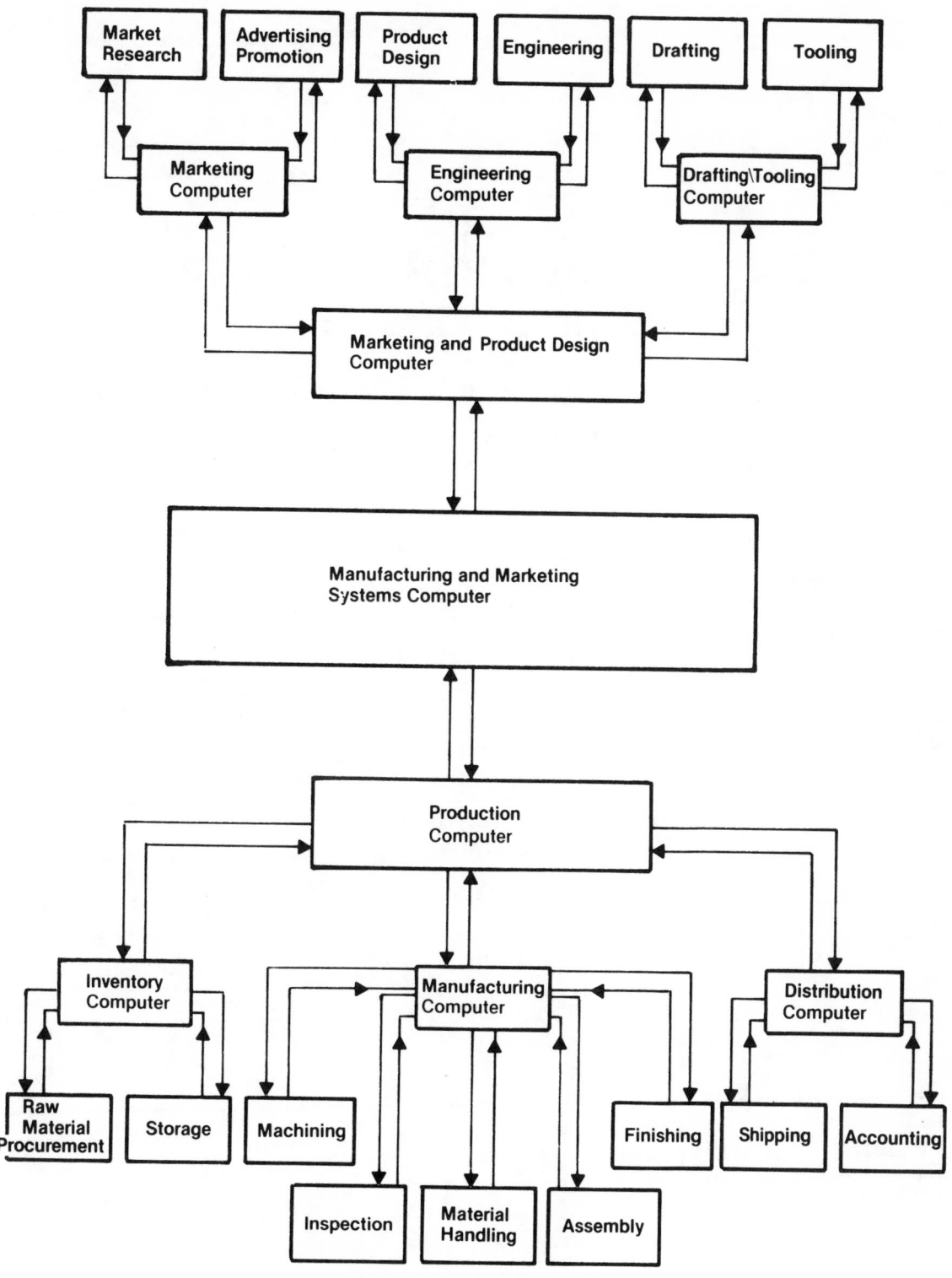

Figure 1 A Typical Schematic for CAM System

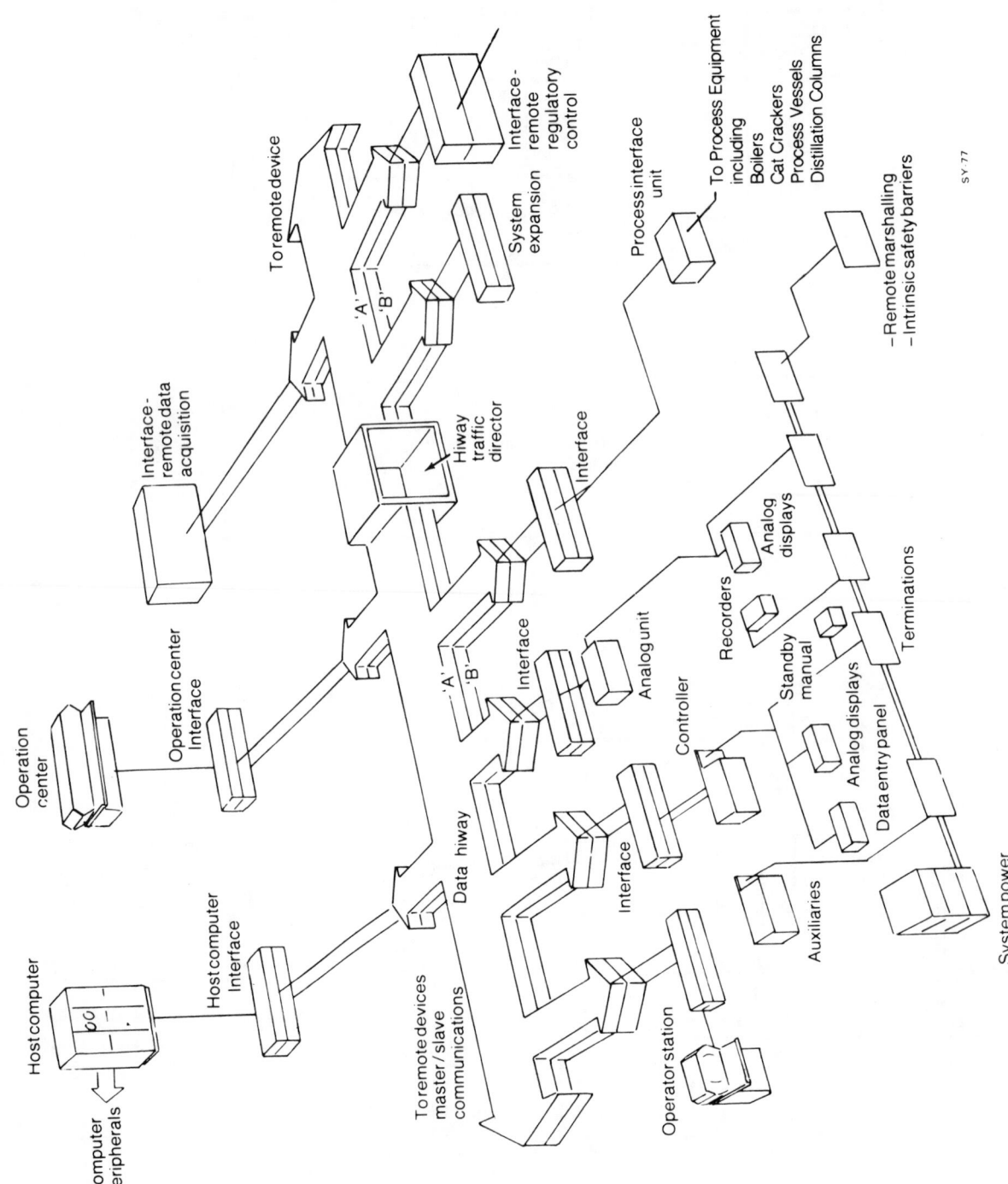

Figure 1B. Typical Schematic for a Complete Distributed Digital Process Control System.

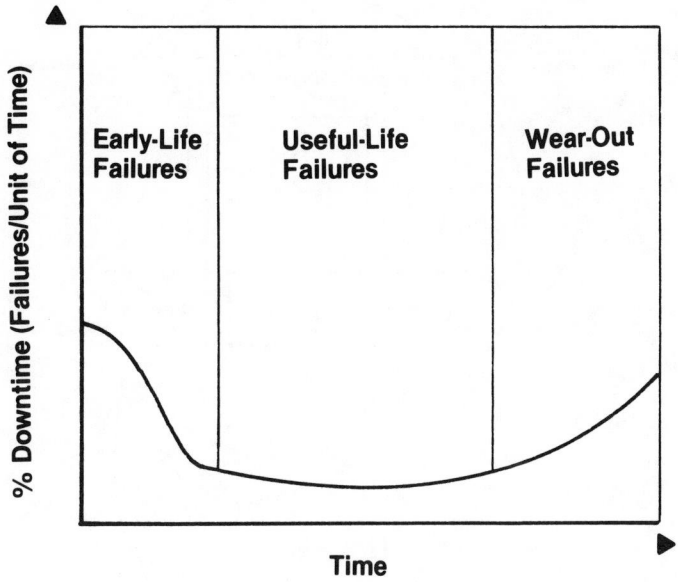

Figure 2. Nature of Failures.

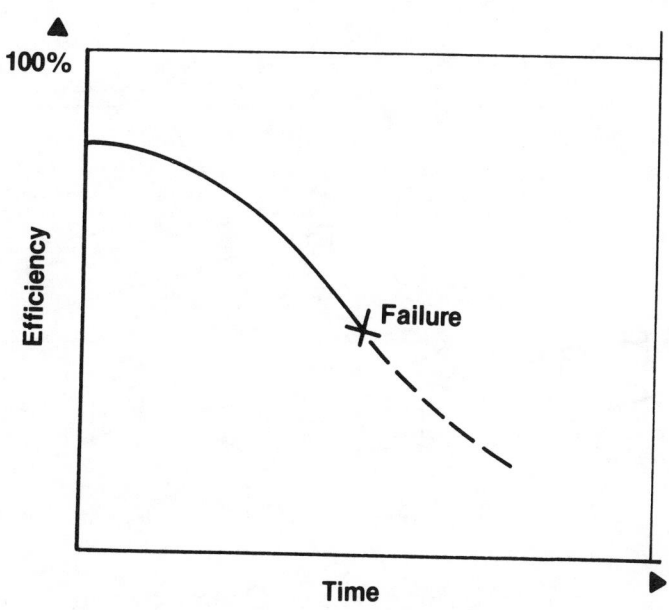

Figure 3. Deterioriation Followed by Failure.

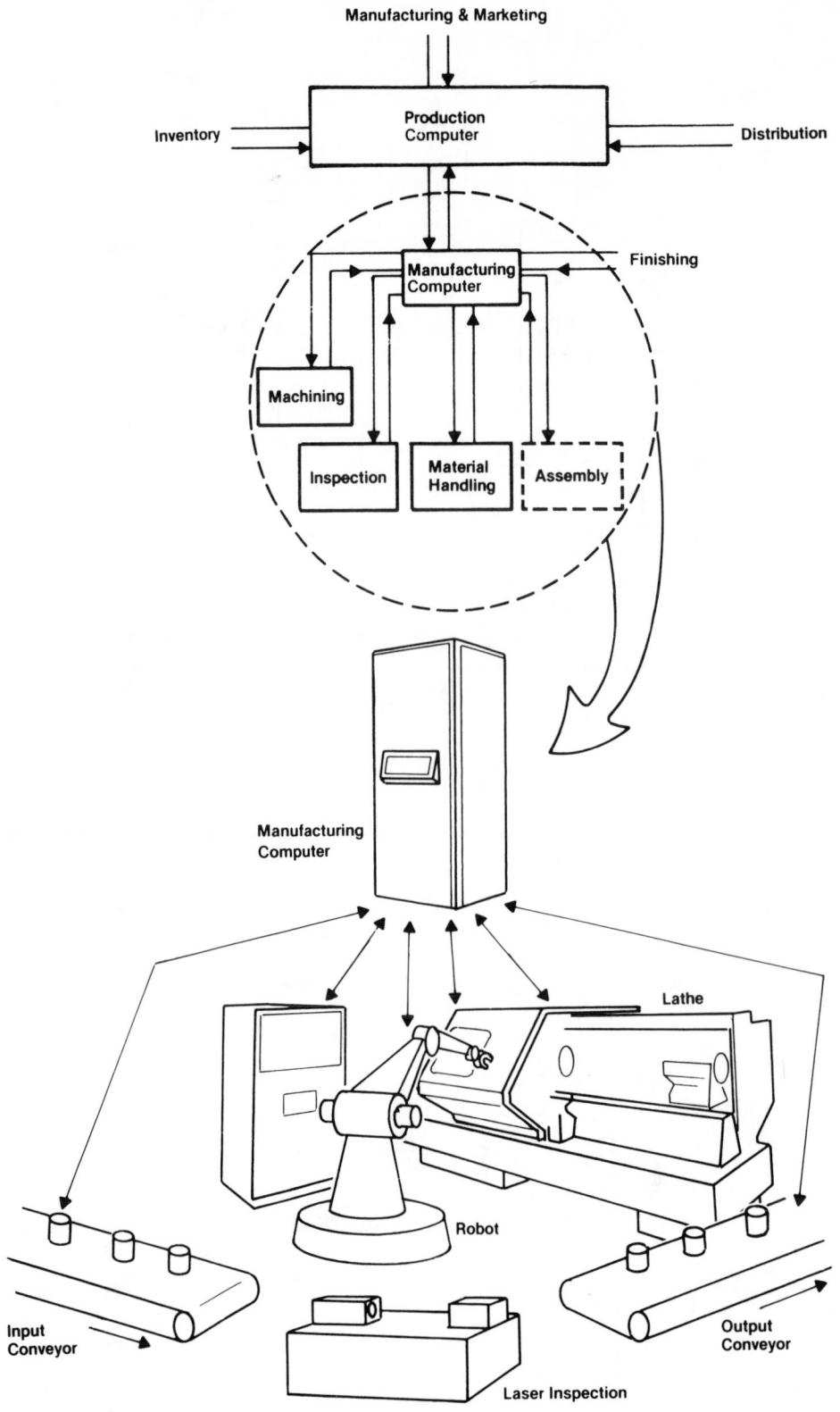

Manufacturing & Marketing

Inventory

Production
Computer

Distribution

Manufacturing
Computer

Finishing

Machining

Inspection

Material
Handling

Assembly

Manufacturing
Computer

Lathe

Robot

Input
Conveyor

Output
Conveyor

Laser Inspection

FIGURE 4 A Key Operation of a CAM System

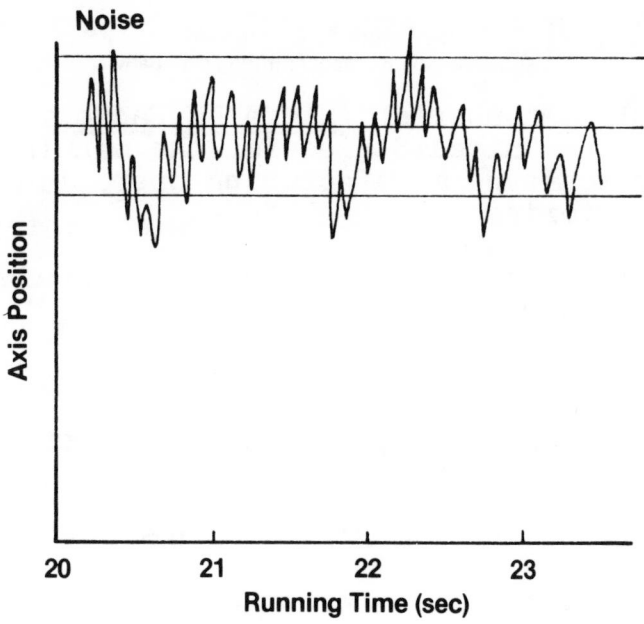

Figure 5. Effect of Noise on System Performance.

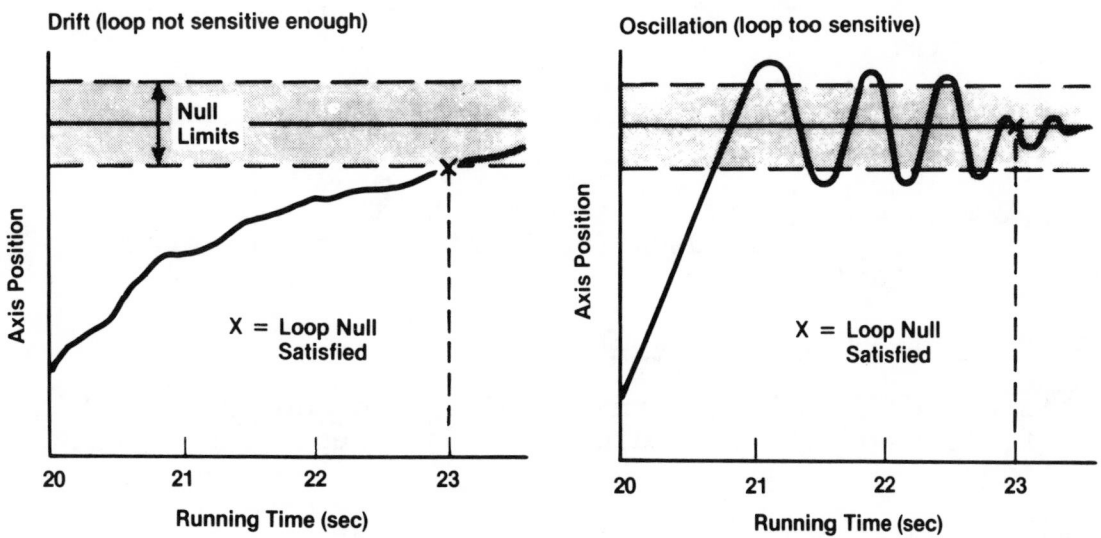

Figure 6. Effect of Drift and Oscillation on Production

Presented at the SME Predictive Maintenance Conference, October 1979

Predictive Maintenance—A Cornerstone for the Automated Factory

By Frank T. Cameron
Brown & Sharpe Manufacturing Company

During the 1980's a number of Automated Factories will be placed in operation in the United States for the manufacture of discreet parts used in batch type manufacturing operations. Unique features of such factories will include --

1. Complete operational control by a Manufacturing Host Computer

2. Zero direct labor

3. Operation on three shifts and seven days per week

4. Operation without manpower on night shifts.

Because of the heavy capital investment, operation of the Automated Factory without "breakdown" is imperative. Essential to the achievement of this objective is an effective and efficient Predictive Maintenance Program.

INTRODUCTION

Would you --

1. start out for Alaska in your car without a spare tire?

2. fly the Atlantic in a plane that had not been serviced?

3. make a deep dive in a submarine that had been sitting in dry dock for twenty years completely unattended?

Of course not and neither would I.

In this context, would you invest 50 million dollars in an automated factory without providing for its effective and efficient operation?

Again, of course not. Predictive Maintenance really begins long before ordering new equipment for your automated factory.

Before considering Predictive Maintenance, let's consider what the Automated Factory for discreet parts manufacturing might look like.

1. There are no engineering drawings.

2. Piece parts are described in geometric terms; are assigned part numbers; and these data constitute part of the data base in the _Engineering Host Computer_.

3. Using a part classification coding system such as that offered by MDSI, all piece parts are classified for prompt retrieval from the _Engineering Host Computer_.

4. A _Manufacturing Host Computer_ (independent from EDP) contains data including --

   a. Complete routing file

   b. The part classification coding system

   c. Complete Production Control capability including --

      Planning

      Scheduling

      Dispatching

      Expediting

      Inventory Control

   d. Total control of manufacturing operations including --

      Machine loading and unloading

      Machine set-up

      In process quality control

      Material handling between work stations

      Storage and retrieval of finished parts.

5. Factory layout will be developed around Group Technology.

6. There will be _zero direct labor_.

7. The _Manufacturing Host Computer_ will be completely interactive with the computer at each work station.

7. a. There will be no tapes.

    b. Complete manufacturing instructions for an
        operation on an order of parts will be trans-
        mitted from the Manufacturing Host Computer
        to the Work Station Computer in one batch.

8. The Automated Factory will operate three shifts and
    seven days per week.

9. A Team of Technicians will work the day shift only.
    The Automated Factory will operate unmanned at night.

10. The Manufacturing Host Computer will have complete
    MIS (Management Information System) capability and
    will be fully interactive with each work station.

11. Machine tools in the Automated Factory will be fully
    depreciated in 3 to 6 years. They will be worn out
    in that period of time.

At least one company in the United States plans to have this
type of Automated Factory in operation during 1980. Using
this concept of an Automated Factory, we may now proceed to
consider Predictive Maintenance.

What do we mean by Predictive Maintenance? Predictive Maintenance may be defined as the work required to be performed which
will enable us to operate an automated factory without breakdown.
Elements to consider when establishing a Predictive Maintenance
Program include --

1. ROI - Objective

2. Power supply and other utilities

3. Machine tools and controls

4. Control network for each manufacturing cell and
    the automated factory

5. Foundations and installation

6. Tooling and accessories

7. Material handling

8. Quality control

9. Organization -
        work required to be performed
        effective staffing.

10. Workpiece characteristics

11. Production planning and scheduling and inventory control

12. Communication network --

    information required

    method of transmission

    destination of information

    required action

    feedback.

13. Predictive Maintenance

    Plan and implement required and scheduled Predictive Maintenance to insure that all of the foregoing systems in the automated factory will operate without breakdown.

Failure or "breakdown" in any of these elements will halt operations and simultaneously ROI turns from positive to negative. Your Maintenance Manager must be fully aware of the effective integration of all these key elements to insure operation without "breakdown". Place him on your team early in the planning phase. You cannot afford a "Three Mile Island" type breakdown in your automated factory!

In his recent book, Management Tasks - Responsibilities - Practices, Peter Drucker draws our attention to the words effectiveness and efficiency.

"Efficiency is concerned with doing things right. Effectiveness is doing the right things."

"Effectiveness is the foundation of success - efficiency is the minimum condition for survival."

No matter how efficient the design engineer makes the V-8 engine, it will probably not survive when gasoline reaches $1.50 to $2.00 per gallon. In this light, our Predictive Maintenance must be both effective and efficient. First, we must decide what is right - what is necessary. Having decided what to do, we must do it well. Some examples should help to sharpen the focus on our approach to Predictive Maintenance.

First, a look at organization. In many companies, factory employees are represented by a union. Typically employee transfers and promotions are based on seniority. At a plant in one large mid-western company, all electricians and electronic technicians were classified in a single seniority group. During a plant expansion in the summer months, the senior electricians were installing new electrical services overhead where it was extremely hot while the electronic technicians were servicing N/C machines on the shop floor. The electricians decided to exercise their "seniority rights" in order to "bump" the technicians to the hot overhead work, and have the opportunity to be in the cooler atmosphere on the factory floor. Even worse, the electricians were not qualified to perform N/C maintenance. Nevertheless, this transfer actually happened in an Iowa plant. Imagine the chaos which would result should this episode be repeated in your automated factory.

Or, consider power supply and installation. Transients and spikes are common in electric power supplied by utilities. Also in our factories, equipment such as welding machines and D/C SCR drives create conditions which assure the failure of N/C controls and computers. Component damage or failure and machine smash-ups are likely results from such transients and spikes. Were this to happen in an automated factory, the cost would be prohibitive --amounting to thousands of dollars per hour. At the time of power distribution design, insure that required isolation, filtration and grounding are provided. Otherwise a major breakdown in your automated factory is predictable.

Now let's turn to diagnostic evaluation of systems and system components. This is an example of the ongoing Predictive Maintenance work to be performed. In any given control and feed-back loop, there are often several components such as a firing card, another P/C board, a resolver, a tachometer, an encoder, etc. Through continual usage over a period of time, one or more of these components within the loop may begin to fail. Common practice today is to await failure; suffer a machine or system breakdown; and identify and replace the defective component. For the automated factory, this practice is unacceptable. Instead, one alternative is to select critical control circuits and periodically diagnose voltage and amperage characteristics of each circuit. With a two-channel recorder such as is offered by Gulton Industries, you can obtain a strip chart of current-voltage data similar to that on an electrocardiogram. Based on periodic analysis of these charts, it is possible to spot potential failures and replace worn or defective components before "breakdown" occurs. Since nobody can afford a "Three Mile Island" type experience in an automated factory, the foregoing approach can prove very effective.

These are but three examples of the kind of work required to develop and implement a Predictive Maintenance Program in your automated factory. Time and space do not permit the development and presentation of a complete program in this paper. In any case, the most effective program will be the one you develop and implement in your own setting. A few guidelines may prove helpful.

1. Learn and state the ROI objective for your automated factory in measurable terms.

2. Develop and record the key elements and/or systems required for <u>effective</u> and **<u>efficient</u>** operation of your automated factory.

   a. The 13 elements previously listed constitute a starting point.

3. For each element or system determine and list the sub-systems that must be operational to insure against breakdown.

   a. Assure yourself that each such sub-system is acceptably reliable.

   b. Develop and implement those maintenance procedures which will insure <u>"breakdown" free operation</u>.

4. A few specific suggestions --

   a. For machines performing critical work, provide a foundation that weighs <u>1 1/2 times</u> the weight of the machine. Isolate the foundation from the factory.

   b. In addition to electrical <u>isolation</u> and <u>filtration</u>, provide a <u>ground</u> of <u>5 ohms</u> or less.

   c. Filter hydraulic and lubricating oil to <u>3 microns</u> or less. (Please refer to American Machinist January, 1978 - Clean Oil Cuts Costs). We have done this at Brown & Sharpe since 1972, and have not replaced these oils in more than <u>seven years</u>.

   d. Develop and maintain an adequate supply of <u>maintenance spares</u>.

   e. Design your communication network to insure that if scheduled maintenance is not performed, the affected machine will be shut down automatically.

   f. Provide for periodic major overhaul and/or rebuilding of critical equipment.

5. Establish an MIS (Management Information System) which presents production performance vs standard for each work center as well as utilization of each work center per time period. Using the "exception principle", reports should reflect deviations beyond acceptable limits. Even though no operator is involved, it is essential to know whether or not the

automated factory is operating to plan. If not - why not?

6. Organization and Staffing. Having determined what work is required, select people who will be effective and efficient. One company that is building an automated factory will have no direct labor workers of any kind. All loading and unloading, set-up and material handling between work centers will be completely automated. Various technicians who service this automated factory, including maintenance people, will be salaried. Even though the factory will operate three shifts and seven days per week, nobody will work on the second or third shifts. This will be a batch type discreet parts making facility.

7. Having determined the critical systems and subsystems in your automated factory, consult fully with the various vendors to develop a complete understanding of all operating characteristics plus vendor recommendations for appropriate maintenance procedures. Include these in your Predictive Maintenance Program.

8. Develop and maintain an adequate supply of maintenance spares. Although this has been previously mentioned, it is important enough to be repeated. Insure against long-term "breakdown" of your automated factory. Consider stocking the following spares which would not normally be stocked for individual N/C machines --

> Spindle drive motors
>
> Axis drive motors
>
> Complete sets of servo valves
>
> Complete CNC controls
>
> Complete sets of roller packs
>
> Complete sets of P/C boards
>
> Selected P/C board components
>
> Transformers
>
> Encoders, tachometers and resolvers
>
> Alignment tools

8. Cont'd.

    SCR drives

    A factory host computer.

9. At the time of equipment purchase, <u>insist on stan-</u>
    <u>dardization</u>. Many procedures including maintenance
    will be simplified. Some examples include --

    CNC controls or micro processors

    Electric motors

    Transformers

    Servo mechanisms

    Pumps.

10. Analyze existing maintenance procedures and records,
    and based on experience, apply what has proved suc-
    cessful. Some examples of such maintenance records
    and procedures are included in the appendix.

Please remember that your automated factory will be unique.
The <u>Predictive Maintenance Program</u> for your automated factory
must be developed to meet these unique characteristics.

SUMMARY

To help you develop your <u>Predictive Maintenance Program</u>, other
speakers on the panel will cover in greater depth some of the
topics outlined in this presentation. Just as the transition
from the old line-shaft operated shops to modern N/C installations required radical changes in organization and operation,
we must anticipate similar drastic changes in transition to
the automated factory. Our success as managers will be measured by the <u>effectiveness</u> and <u>efficiency</u> with which we
achieve transition to the automated factory. Everyone present
will see the impact of the automated factory in his business
<u>by 1985</u> - probably even sooner. In summary, our <u>Predictive</u>
<u>Maintenance Program</u> can be effective if we focus clearly on
key points such as --

1. Know your <u>ROI objective</u>.

2. Your Automated Factory must operate <u>without</u>
   <u>breakdown</u>.

3. Your <u>Maintenance Technicians</u> will be part of
   the <u>Technician Team</u> who operate your Automated

3. Cont'd.

   Factory.

4. The <u>Manufacturing Host Computer</u> will be the nucleus - the nerve center - of your Automated Factory.

5. Your Automated Factory will operate <u>three shifts</u> and <u>seven days per week</u> with no night shift personnel.

Quite a challenge we face during the next decade! Imagine the Automated Factory to be like a three-legged stool. <u>One of those legs is your Predictive Maintenance Program - remember, no breakdowns! You cannot sit properly on a tilted stool.</u>

APPENDIX

1. Maintenance Work Order Procedure

2. Oil Filtration Procedure
&
3.

4. Weekly N/C Machine Performance Report

5. P/M Work for W&S 2-SC N/C Turning Machines

Brown & Sharpe Mfg. Co.
Maintenance Work Order Procedure 1.1.3.8

The purpose of this procedure is to provide an effective and economical means for planning, scheduling, and dispatching maintenance work. Maintenance Work Order Form No. 821-1664 will be used with this procedure.

This procedure will apply where the cost of maintenance work required does not exceed fifty (50) hours for labor or $500 for material. When either of these limits will be exceeded, the Work Order Form 457 will be initiated and approved per present practice.

Any foreman or department supervisor requiring maintenance work should initiate Maintenance Work Order Form No. 821-1664. The information to be entered by the originator is shown on the following sample:

TIME NUMBER OR EXPENSE ACCOUNT #		ORDER WRITTEN	
		TIME	DATE
5334-317-316-12		10:30 A.M.	8/1/79
DEPT. / SECTION	LOCATION	DESCRIPTION OF MAINTENANCE	
		REPAIR DEFECTIVE CLUTCH ON MAIN DRIVE.	
EXPECTED COMPLETION DATE			
8/3/79			
ACTUAL COMPLETION DATE			
TIME	DATE		

B & S

MAINTENANCE WORK

ORDER FORM

SIGNATURE

N. E. SMITH

FORM 821-1664

The originator will retain the original (white paper copy) and deliver the remaining two copies to the Maintenance Department.

Maintenance Work Orders involving installation of new equipment, relocation of existing equipment, and/or removal of old equipment will originate in the Plant Engineering Office. These Maintenance Work Orders will result from receipt of Machinery and Equipment Installation or Transfer Slips (Form 278) from the appropriate Division Director of Manufacturing Engineering or his designated representative. It is requested that the following lead time schedule be used in the issuance of these Transfer Slips:

Brown & Sharpe Mfg. Co.
Preventive Maintenance Filtration
of Lubricating and Hydraulic Oils
in Production Machines
Procedure 1.1.3.12

The purpose of this procedure is to establish the method and
frequency for filtration of lubricating and hydraulic oils on
designated critical production machines at Precision Park.

Equipment presently used for this filtration work is the IMP-356
(Imperial Hydraulics) equipped with a Brown & Sharpe pump and
having a filtration capacity of 5 gallons per hour. This is a
small, portable, electrically operated unit with three-stage
filtration. These filters which are disposable are arranged in
series as follows:

                    1st. stage - 100 microns

                    2nd. stage - 40 microns

                    3rd. stage - 10 microns

                    4th. stage -   3 microns

1. N/C Machines will be scheduled for monthly filtation of oil.

2. Other machines listed in Schedule A will be scheduled for bi-
    monthly filtration of oil. (Once every two months)

3. For each 5 gallons of oil reservoir capacity, the filtration
    unit will be run for one hour.

The objective of this procedure is two-fold:

1. By predetermined oil filtration schedules, keep lubricating
    and hydraulic oils clean and acid free and thereby extend
    the oil life to 5 years! (Past practice has been to change
    oil in critical machines yearly.)

2. By predetermined oil filtration schedules, keep lubricating
    and hydraulic oils clean and acid free -- thereby avoiding
    expensive breakdowns.

Procedure and Frequency

1. The Maintenance Services Engineer in cooperation with the
    Department Manager and the Maintenance Foreman will select
    and list critical production machines in which hydraulic
    and lubricating oils shall be filtered.

Brown & Sharpe Mfg. Co.
(Enclosure A) Procedure 1.1.3.12

NUMERICALLY CONTROLLED MACHINES

Dept. No.	Mach. No.	Machine Description	Cap. In Gals.	Filter. Time (Hrs.)	B&S Oil Type	Oil Type	Loc.
Machine Tool Division							
5335	13-68	B&S Hydromaster	42	9	#53	Sunvis 701	N-23
5335	13-69	Sundstrand OM80 Hydraulic ---	50	10	#62	Sunvis 706	N-23
		Mach. Lube System ---	50	10	#81	Sunvis 500	N-23
5335	13-70	Sundstrand OM80 Hydraulic ---	50	10	#62	Sunvis 706	N-23
		Mach. Lube System ---	50	10	#81	Sunvis 500	N-23
5335	13-71	Hydromaster	42	9	#53	Sunvis 701	O-23
5334	14-102	Behrens Press	200	40	#64	Sunvis 747	O-30

Downtime Report 1979

W&S 2SC 7-518	Week 3/24 Hrs.	%	Week 3/31 Hrs.	%	Week 4/07 Hrs.	%	Week 4/14 Hrs.	%	Week 4/21 Hrs.	%
Hrs. Sched.	164.5		162.5		164.5		117.5		141.0	
Hrs. Run	150.3	91.4%	154.0	94.8%	160.9	97.8%	114.5	97.4%	124.2	88.1%
Maint. Down	5.5	3.3%	1.0	0.6%	1.5	0.9%				
Prod. Ctl. Down										
Mfg. Eng. Down	0.7	0.4%					3.0	2.6%	12.3	8.7%
Inspec. Down										
Tools Down										
Dept. Down	8.0	4.9%	7.5	4.6%	2.1	1.3%			4.5	3.2%

W&S 2SC 7-519	Week 3/24 Hrs.	%	Week 3/31 Hrs.	%	Week 4/07 Hrs.	%	Week 4/14 Hrs.	%	Week 4/21 Hrs.	%
Hrs. Sched.	164.5		162.5		164.5		117.5		141.0	
Hrs. Run	159.5	97.0%	152.0	93.6%	141.0	85.7%	117.5	100.0%	136.6	96.9%
Maint. Down			7.0	4.3%	9.0	5.5%				
Prod. Ctl. Down										
Mfg. Eng. Down	1.0	0.6%				12.4%			2.0	1.4%
Inspec. Down										
Tools Down			2.0	1.2%						
Dept. Down	4.0	2.4%	1.5	0.9%	2.1	1.3%			2.4	1.7%

Brown & Sharpe Mfg. Co.
Weekly N/C Machine Performance Report

Brown & Sharpe Mfg. Co.
P/M Work for W&S 2-SC
N/C Turning Machines

The following maintenance intervals are suggested for use under optimum conditions. If environmental conditions warrant, the frequency of maintenance intervals should be increased.

Frequency	Service "Electrical"
Weekly	
Monthly	Air Filters: Air filters are located on the machine control cabinet. N/C control cabinet. Longitudinal servo drive and D.C. main drive motors. Inspect and clean the air filters with hot water and a detergent or any type of solvent and then blow them dry with air.
Monthly	Control Panel:

Control Panel:

1. Check for dirt, oil or water in control panel area and clean.

2. Check for airtight seal of control panel area.

3. Check tightness of screws on terminal boards and relays.

4. Check all switches and operating buttons including signal lights.

3 Month — D.C Main Motor; Servo Axis Motors; Index Drive Motors and D.C. Tachometers

1. Clean out any accumulated dirt and dust.

2. Check brushes.

3. Check brush spring tension.

Reprinted from *National Safety News*, December 1980

Materials Handling Equipment
Preventive and Predictive Maintenance

By Frank V. Claire, *Division Vice-President, Management Consulting Division, The Austin Company, Evanston, IL.*

IN MOST MANUFACTURING PLANTS today the concept of preventive and predictive maintenance (PPM) is the subject of much discussion. In many cases, unfortunately, that discussion is the only action taken. In other cases, there is a relatively scattered application of preventive maintenance efforts on production machines, but with little or no consideration of the needs and importance of the materials handling equipment.

The existence of either of these two cases in the area of your materials handling equipment indicates a strong need for the application of a systematic approach to the justification and allocation of PPM efforts to the complete facility equipment complement, including the materials handling equipment.

Ben Franklin wisely pointed out, "An ounce of prevention is worth a pound of cure." When it comes to maintaining essential equipment, this is especially true, because the "pound of cure" generally means emergency repair of a critical breakdown. This corrective maintenance often costs more than just parts and labor. Failure of one part of a machine can lead to damage of other parts, resulting in further maintenance problems. Also, if the equipment is especially critical to operations, such as air temperature control, utilities services, or production and materials handling machinery, the breakdown will likely lead to more serious consequential costs; for example, having to shift the work force to non-productive tasks.

EDITOR'S NOTE: This article is based upon a presentation made by Mr. Claire at the Materials Handling Institute Forum held at Chicago on Sept. 10, 1980.

The proper care of normally operating equipment is preventive maintenance. The lack of it is neglect, and will lead to premature equipment failure.

Predictive maintenance is based on the fact that most equipment failures are preceded by detectable indications that failure is imminent. These indications take a variety of forms—increased vibrations, temperature rise, increased power consumption, and so forth. By monitoring and measuring these changes, management can predict failure and act before an untimely breakdown occurs.

Preventive maintenance extends the useful life of equipment, and predictive maintenance allows management to anticipate equipment failure and to schedule corrective work on a timely basis. Combined, preventive and predictive maintenance serve as an ounce of prevention to help avert the costs incurred with unexpected breakdowns.

Maintenance and Money

Many managers fail to give proper consideration to the costs of machine downtime. They accept unnecessary emergency machine outages as a normal cost of doing business. In today's economic climate, with its high labor and high capital equipment cost, responsible managers cannot accept this drain on profit.

The costs associated with machine maintenance can be separated into three components:

• *Out-of-service cost*—Resulting from emergency equipment outages, it is the cost of whatever action management undertakes to compensate for lost operations and production. This includes scheduling high-cost overtime work, subcontracting production at higher than in-house cost, and even making additional capital expenditures to provide backup capacity.

• *Repair cost*—This covers the parts and labor required to restore equipment to use following a breakdown. It also includes the cost of increased labor and material budgeted to combat excessively high cost of emergency equipment failure.

• *PPM cost*—For systematic monitoring and servicing of equipment, which not only reduces the frequency of failure, but also minimizes the consequences of equipment failure.

Combined, these three costs represent the total controllable maintenance cost of an operation. An effective PPM program will reduce these costs to a minimum range, achieved by reducing equipment out-of-service costs and repair costs. These costs will decrease as PPM costs increase—but only to a degree. An effective PPM program manages expenditures in the minimum cost range on selected equipment, which shows economic justification for being in the program.

Creating a PPM Program

To design an effective materials handling equipment PPM program, three major questions must be answered to ensure the greatest economic return for the PPM expenditure:

• Which equipment should receive PPM?

• What kinds of PPM should be performed?

• How often should PPM be done?

Unless these questions have been addressed in a systematic and logical manner, no PPM program can provide maximum benefit for the monies expended. For every equipment item, the cost of appropriate PPM measures must be compared to achievable reductions in repair costs and out-of-service costs. Only then can controllable maintenance costs be minimized.

In order to make an accurate determination of the overall PPM needs, it is necessary to conduct a complete physical audit of all equipment, including production equipment and materials handling equipment. The audit must provide the following information for each equipment item:

• Identification by name, inventory number, location;

• Level of criticality to operations;

• Equipment category by function performed;

• Equipment class and complexity;

• Estimated replacement cost;

• Potential out-of-service loss per failure;

• Estimated repair costs;

• Operating conditions and environment;

• Annual hours of operation.

A simple form must be devised for the listing of such an equipment audit.

Then, through the use of computerized analysis, the large mass of data accumulated in the equipment audit can be processed and evaluated in relative terms, by comparing the findings on each piece of equipment in the facility with all other facility equipment. This approach results in a calculation of the amount of PPM effort economically justifiable for materials handling equipment as well as other listed machines.

PPM Techniques

The preceding calculation permits a cost-importance ranking of those equipment items that should be included in a formal PPM program. Normally, it is found that somewhere between 15 and 30 per cent of the total number of equipment items in

a facility are properly included in the program.

After the equipment items have been selected, the next step is to make a technical analysis of the PPM techniques necessary to increase equipment reliability, and thereby reduce equipment outages. The purpose of the technical analysis is to identify from a mechanical viewpoint the specific types of PPM techniques that will be most effective in monitoring equipment condition and detecting imminent equipment failure.

This technical analysis provides a means to record each PPM technique required for each equipment item, the importance rating of each technique, and the estimated time to do the work. It also facilitates organizing this information for computer input.

PPM Costs and Savings

The information now available within the computer permits the calculation of both the PPM program costs and the amount of savings that can result from proper application of specified PPM efforts. Such PPM techniques might include: the scheduling of visual examinations and inspections; cleaning; lubrication; check-out of electrical, electronic panels and numerical controls; analysis of vibration, sound, chemicals, and oil; temperature and pressure records; flowage; check of infra-red and other types of scanners; ultra-sonics, radiography, and all other factors that were fed into the computer.

A PPM requirement study of the type under discussion here must consider not only the economic justifications mentioned, but also the various other elements that are necessary in the implementation of a comprehensive program. These include analyzing the maintenance organization and recommending organizational changes, which may be necessary to permit the PPM function to operate as an entirely separate entity from the normal repair function.

Also important is the investigation of the systems procedures and records, which are presently available, and the determination

of any changes and revisions that may be necessary to support a comprehensive PPM program. In most cases, it is also necessary to determine what facilities and equipment, are available to the PPM group, and to identify additional PPM instruments and devices needed to make the PPM measurements.

The implementation of an effective program also will require the development of work schedules designed to maintain an even work load on a day-to-day or week-to-week basis. This will ensure maximum utilization of PPM manpower.

PPM Program Results

The total benefits from a properly designed PPM program can be most rewarding. In an actual case history involving only the production machines in a large machine shop, this approach to justifying PPM efforts resulted in the following information:

• A total of 332 equipment items were audited;

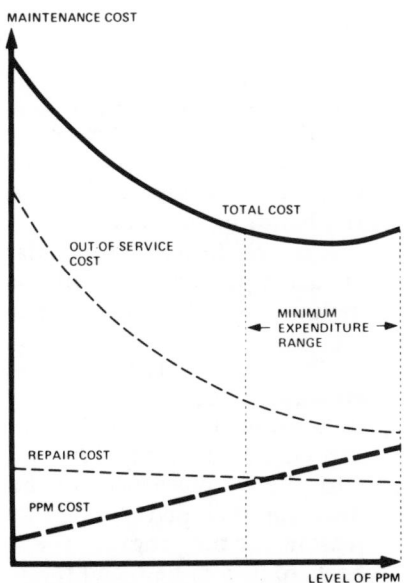

Chart illustrates relationship of out-of-service, repair, and PPM costs, representing total controllable maintenance costs of an operation. An effective PPM program will reduce total controllable maintenance costs to a minimum range, by reducing equipment out-of-service costs and repair costs. As shown, these costs decrease as PPM costs increase, but only to a degree.

- A total of 79 equipment items were identified as economically responsive to PPM efforts;

- Total projected net annual savings through reduced emergency repair, reduced production loss, and extended equipment life amounted to $347,380;

- Selecting a total of four from a crew of approximately 45 employees to maintain the PPM program amounted to an annual labor cost of $135,800.

As is seen in this case history, the marginal return on PPM expenditures can be 2.5 to one, and in many cases even higher. The inclusion of materials handling equipment in this study would surely have changed the allocation of PPM expenditure, and provided improved reliability in material handling.

Thus, this approach gives management the cost information necessary to assess the value of establishing a comprehensive PPM program, and to direct and control maintenance from a cost-justified viewpoint.

Program Implementation

Upon completion of the PPM requirement studies, analysis of the benefits available will quantify the need to develop and install an effective PPM program, including consideration of material handling equipment.

A typical format for installation of such a program normally will include the following elements:

- *Introductory meetings*—These are held to acquaint management staff, production supervision, and maintenance supervision with the various aspects of the program. Included will be the reasons for the program, the goals to be attained, and the manner in which these goals will be pursued;

- *PPM facilities and equipment*—This element includes the determination of specific types and quantities of specialized PPM equipment needed, and the selection and procurement of these items;

- *Selection of PPM personnel*—Includes testing and interviewing candidates selected from the maintenance organization;

- *Systems revisions*—Existing PPM systems, records, and forms are revised as needed to provide a record and control system capable of supporting and monitoring the PPM operation;

- *PPM personnel training*—A series of training programs for the use of PPM equipment and method of operation under the PPM procedures is provided to the selected PPM personnel;

- *PPM work schedules*—The computerized scheduling program is activated and the PPM supervisor trained in the assignment of work and identification of work completion;

- *Control report*—A reporting function of the PPM group is developed and activated to provide management with an on-going review of PPM efforts and results;

- *Program supervision and follow-up*—Program supervision must be provided during the start-up portion of the program to ensure conformance to the original procedures and goals. Also required is intermittent program follow-up and review to correct deviations over a period of about six months to one year, or until the program becomes self-supporting.

Conclusion

Preventive and predictive maintenance programs are applicable to a wide variety of situations, including manufacturing facilities, utilities, hospitals, high-rise office buildings, and transportation facilities. It has been our experience that typical PPM programs can achieve the following results:

- A 25 per cent reduction in the level of emergency breakdowns;

- A 10 per cent extension in equipment life;

- Five to 15 per cent reduction in overall maintenance costs;

- A 20 per cent reduction in equipment downtime.

Perhaps almost as important as the return on investment from a PPM program applied only to one segment of a plant is the fact that this approach provides detailed cost information. This often is necessary to convince top management of the importance of establishing a comprehensive PPM program including the facility's materials handling equipment, thereby reducing the production interruptions caused by unexpected failures in such equipment. Ω

BIBLIOGRAPHY

"Fire Protection for Combustible Metals," (Data Sheet I-567-79). NATIONAL SAFETY NEWS, 444 N. Michigan Ave., Chicago 60611. Vol. 119, No. 6, June 1979, pp. 75-82.

"Spotlight on Energy Conservation, Safety, and Health Innovations at Trade Shows." NATIONAL SAFETY NEWS, 444 N. Michigan Ave., Chicago 60611. Vol. 120, No. 1, July 1979, pp. 75-79.

"Unloading Bulk Grain from Freight Cars," (Data Sheet I-521-79). NATIONAL SAFETY NEWS, 444 N. Michigan Ave., Chicago 60611. Vol. 120, No. 6, December 1979, p. 63-67.

"Do Your Storage Racks Stack Up With Industry Standards?" NATIONAL SAFETY NEWS, 444 N. Michigan Ave., Chicago 60611. Vol. 120, No. 6, December 1979, p. 71-73.

"How to Get the Most Out of Your MHE (Materials Handling Equipment)," NATIONAL SAFETY NEWS, 444 N. Michigan Ave., Chicago 60611. Vol. 120, No. 6. December 1979, pp. 74-75.

"This M/H Job Was Tough—But Oh, So Gentle." NATIONAL SAFETY NEWS, 444 N. Michigan Ave., Chicago 60611. Vol. 120, No. 6. December 1979, p. 78.

Frank V. Claire holds a B.S. degree in Mechanical Engineering, with an Industrial Engineering option from Purdue University. His fields of concentration are in production and operations, including labor controls, methods engineering, facility planning, and production and inventory control. He is a founding member of the Institute of Consulting Engineers, and also is a senior member of AIIE, APICS, SME, and AIPE.

Unlocking Captive Production Time With Control Failure Knowledge

By DONALD E. HEGLAND
Associate Editor

Dollars lost to downtime bite a big chunk out of U.S. industrial manufacturing profits every year. Downtime cost the average plant about $260,000 in 1978, according to a survey of PRODUCTION ENGI-NEERING readers at that time. And in a few very large plants, readers reported downtime costs as high as $6 million a year.

Downtime is a major negative force, slowing production and hampering efforts to boost productivity in the U.S. Production equipment can be knocked out of business in many ways, of course, but one of the more significant mechanisms is by failure of a control system or component on which the equipment depends for power management, guidance, or any one of innumerable other functions.

Accordingly, a major survey of PRODUCTION ENGINEERING readers sought to learn what types of control components are most often responsible—by their failure—for causing downtime. And—more importantly—to address several related questions:

■ What are the most common control component failure modes?
■ What problems are most often encountered in restoring failed control components to service?
■ How much does control-

The true cost of downtime

When production equipment stops producing, the cost to the manufacturing enterprise includes both the cost of the necessary repairs and the loss of that which would have been produced. We asked PRODUCTION ENGINEERING readers to estimate how many hours of production would be lost if certain control components failed, and to estimate both the dollar value of those lost production hours and the direct cost of materials and labor for repairs.

Twenty categories of control components, ranging from ac and dc adjustable-speed drive controls to the ubiquitous limit switch, were represented on the "what if this fails" list. And although the numbers vary widely, they bear out what we all know intuitively—that we have an expensive downtime problem in U.S. manufacturing. Time lost estimates varied from a few hours to nine days, with nearly half a shift of lost production being about average. And the value of lost production was pegged everywhere from a few hundred dollars to $24,000. The ratio of lost-production costs to repair costs was from 2:1 up to 15:1, with 4:1 being about average. In only a few cases did respondents estimate that repairs would cost more than the lost production. Here's a brief summary of some of the responses.

■ Not surprisingly, many respondents said their production operations are quite vulnerable to failure of adjustable-speed drive controls, both ac and dc. Estimates of lost production time due to downtime of these devices ranged from a few hours up to over a week, with four to eight-hour estimates common and one or two days not unusual.

Lost production dollar estimates ran from a few hundred dollars up, with one respondent quoting $5,000/h, and figures of several thousand dollars common. Repairs are costly too; estimates ranged from a few hundred to a few thousand dollars, with $500 about average.

■ Motor starters, and control centers, were also big-dollar items in terms of downtime costs. Respondents estimated lost production times from one hour to three weeks, with the average about one full shift. And the resultant production loss estimates were in much the same range as those for drive controls; from $100 to 250,000, with several thousand mentioned frequently. Repair costs, too, were in the same ballpark as those for drive controls; from $50 to 10,000, with the average near $500.

■ Even the smaller components—switches, relays, transducers, counters, photoelectrics, fuses, recorders, circuit breakers, etc.—have the potential for bringing operations to an expensive halt. How expensive depends, of course, on how critical the component is to the operation. Many respondents estimated losses in production time of up to a full shift or more due to failure of any one of these devices. And they said these periods of downtime frequently rang up losses in the thousands of dollars. The only consolation was that repair costs tended toward the low side, averaging under $100.

■ The responses indicated that numerical controls and programmable controllers are less often the cause of downtime than, for example, adjustable-speed drive controls, which probably reflects the greater number of the latter in use more than anything else. But NC and PCs also show up as responsible for expensive downtime, probably because they are typically applied to high value-added processes where interrupted production tumbles profits in a hurry.

Respondents estimated from four hours to six weeks of lost production time from these devices, with the average about one full shift. This downtime translated into losses estimated from several hundred dollars up to $40,000, averaging about $1,000. And estimates of repair costs ranged from $100 to 5,000, averaging about $400.

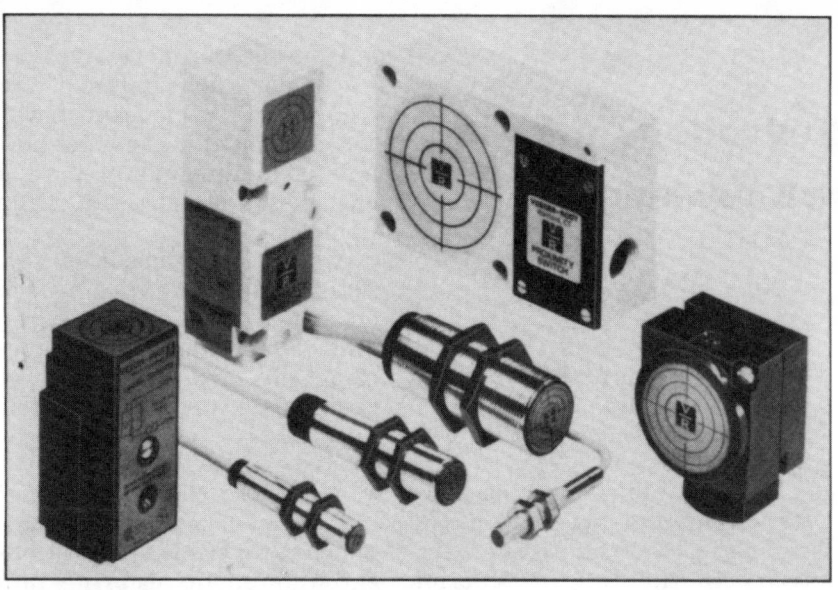

Proximity switches replace mechanical limit switches

Line of self-contained inductive proximity switches from Veeder-Root includes direct replacements for industrial mechanical limit switches, modular plug-in models for ease of installation and replacement, and tubular models with diameters from 8 to 30 mm. Fixed, nominal sensing spans range from 1 to 40 mm. One series has an adjustable sensing span of 30 to 60 mm, and another series contains an internal timer. Rugged solid-state design is said to make them ideal for industrial environments where fluids, dirt, shock, and vibration are present.

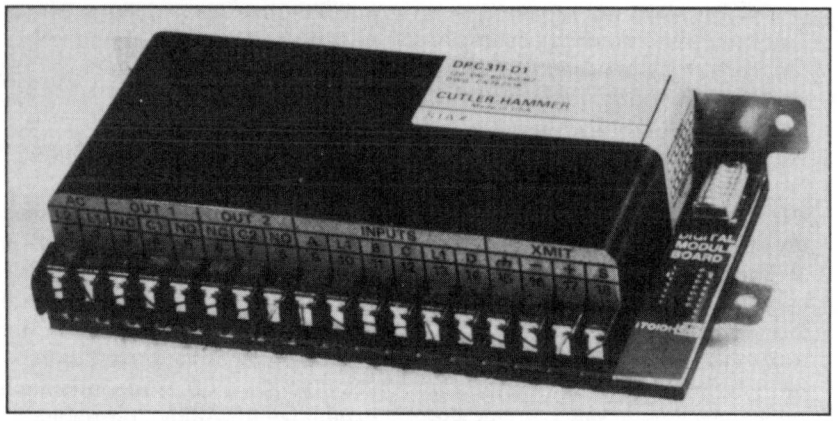

Control station is for harsh industrial environments

Microprocessor-based control station is designed for inexpensive signal collection and output control in harsh industrial environments. Cutler-Hammer DPC311 stations, by Eaton Corp.'s Market Development Div., indicate status; monitor control signals and temperature; and control, pulse, and latch relay outputs based on control command. All control communications are accomplished over inexpensive two-wire shielded cable to reduce overall cost of distributed intelligent monitoring and control. Up to 127 stations can be tied to the two-wire cable and one computer interface adapter can drive two serial cables. Each station has a power supply to assure reliability.

failure-generated downtime cost, both in terms of production lost forever and in repair bills?
■ What kind of downtime-reduction programs are production engineers carrying out today?
■ What would they like to see control component vendors do to help these efforts?
■ How well are company managers doing in supporting the production engineering function's downtime-reduction programs?

Bar charts summarizing the data derived from the first three questions can be found throughout this article. And responses to the last three are discussed in the remainder of the text.

Preventive maintenance wins

Readers were asked to list the two or three most important (in their opinion) things that they do now to reduce downtime due to control failures, in their manufacturing operations. Not too surprisingly, 66% of the respondents reported planned preventive maintenance programs already in place. And another 57% said that they maintain some inventories of spare parts, ranging from just a few critical items to fully stocked spare parts shelves. At the same time, many indicated dissatisfaction with the capital appropriations they could get for these two activities. More about this later.

A number of different approaches to reaching the same goal—maximum productivity with maximum uptime—were mentioned. The importance of good training programs was cited frequently, and 23% of the respondents reported that they train maintenance people either in formal in-house programs or at vendor training schools.

Another 7% pointed out that good downtime-reduction programs begin with buying good equipment, and not necessarily going with the lowest bidder. Here, too, many cited difficulty in implementing this philosophy due to reluctance on the part of management to approve the necessarily higher funding. Others point to pre-installation shakedowns and insistence on obtaining self-diagnostics as important parts of their anti-downtime campaigns.

Planning for the necessary maintenance at the project concept stage, before equipment is purchased, was mentioned by one respondent as effective in minimizing downtime later on. And several others mentioned things like systems redundancy, parallel operations capable of exchanging overloads caused by downtime, and spare capacity in each of several operations to allow sharing the work from a line that is out of service.

We need spares and documentation

Readers were also asked what steps they think the manufacturers of control components and systems should take to minimize the downtime caused by their products. And 22% of the respondents called immediately for better maintenance documentation; some even said *any* documenta-

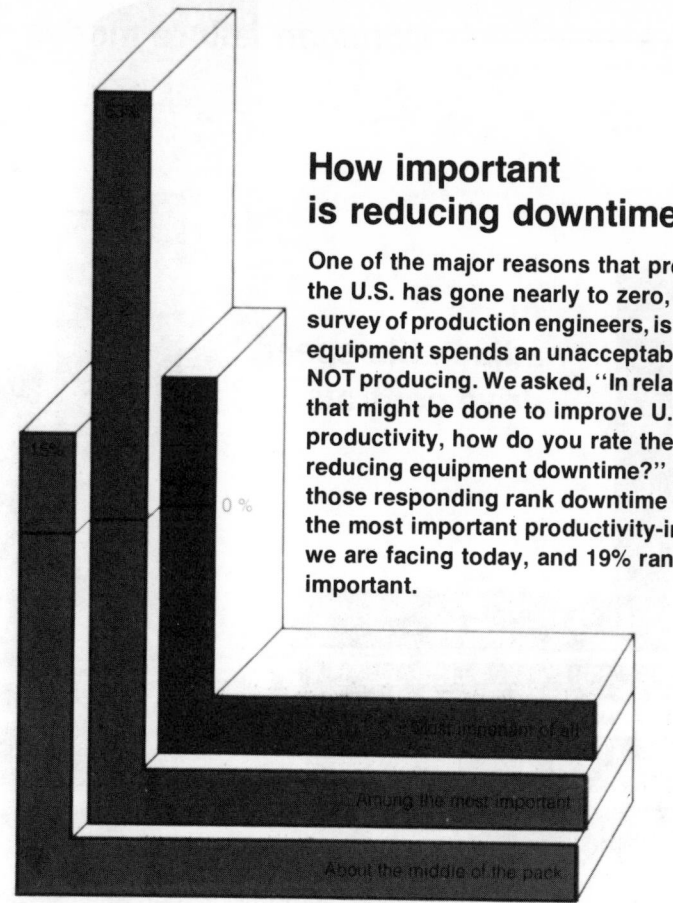

How important is reducing downtime?

One of the major reasons that productivity growth in the U.S. has gone nearly to zero, according to our survey of production engineers, is that our production equipment spends an unacceptable portion of its time NOT producing. We asked, "In relation to all the things that might be done to improve U.S. manufacturing productivity, how do you rate the importance of reducing equipment downtime?" Nearly two-thirds of those responding rank downtime reduction as one of the most important productivity-improvement tasks we are facing today, and 19% rank it THE most important.

Which controls cause downtime?

They all can—and do—of course. We asked a number of PRODUCTION ENGINEERING readers to pick the control components most often responsible, by their failure,

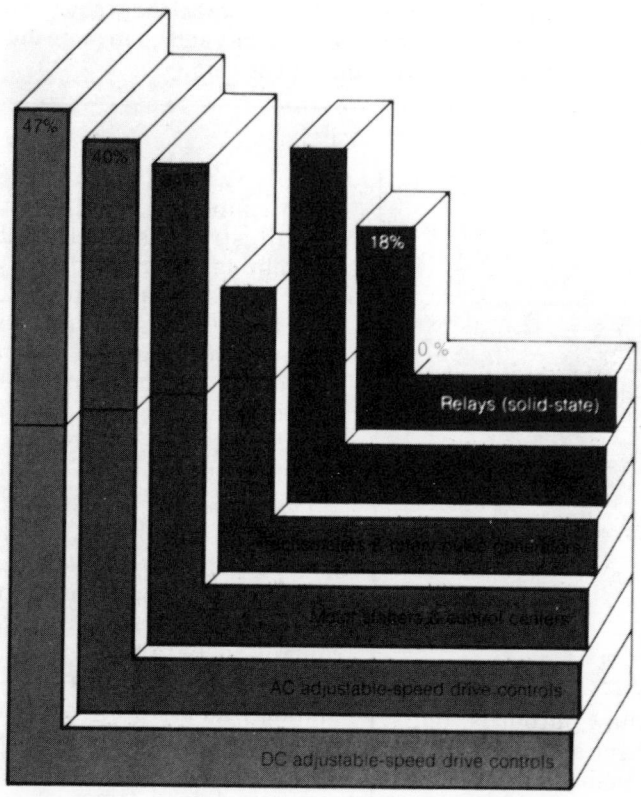

for production-stopping downtime in their plants. The list of 20 control component categories included devices like ac and dc adjustable-speed drive controls, numerical controls, programmable controllers, counters, recorders, motor starters, relays, transducers, switches, circuit breakers and fuses, and others.

These data represent the range of the responses. Tachometers *et al* were tied with optical scanning systems and rotary position encoders for the least often responsible for downtime position: only 13% of the respondents picked them as the worst offenders in their plants. And, the fact that 40% and 47% of the respondents picked ac and dc adjustable-speed drive controls, respectively, as the worst offenders in their plants certainly reflects the heavy concentration of these devices in industry today.

Other control components cited by respondents as most often responsible for downtime in their plants included photoelectric controls (39%); counters (36%); limit, pressure, and proximity-sensing switches (38%); numerical controls for machine tools (31%); fuses and circuit breakers (26%); programmable controllers (24%); and transducers (15%).

We also asked readers to tell us what were the most common failure modes for each control component category, and also the most troublesome aspects they experienced in restoring failed control components to service. Selected data derived from these questions are presented in other bar charts elsewhere in this article.

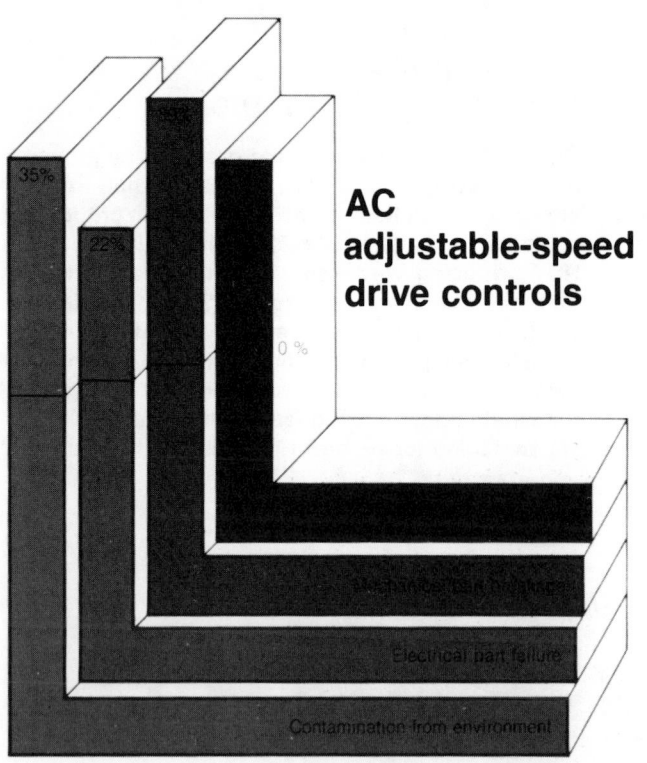

AC adjustable-speed drive controls

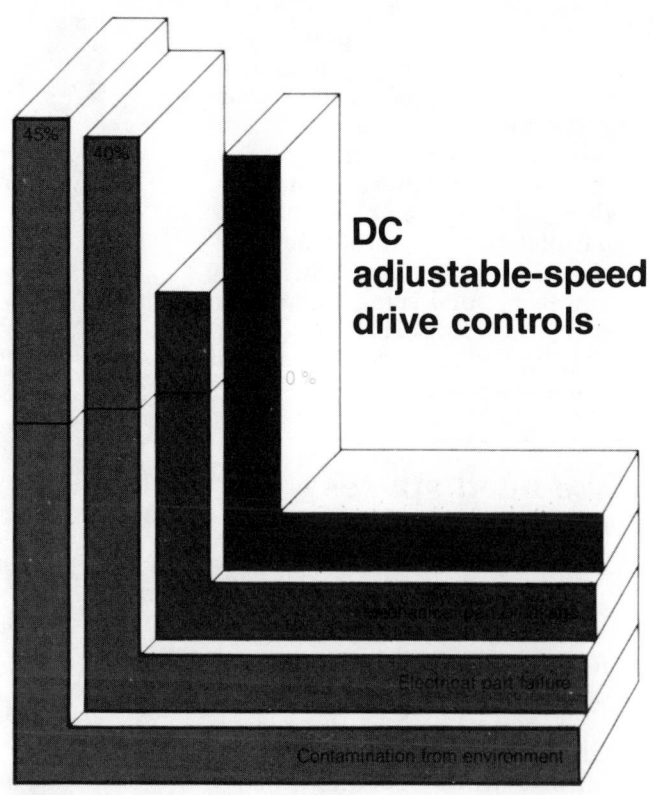

DC adjustable-speed drive controls

Mechanical part breakage puts an ac drive control out of business more than twice as often as it does a dc drive control, according to those responding to our survey, while the dc unit is nearly twice as susceptible to electrical part failure. They're about equally vulnerable to burnout or overheating, it seems, but the dc unit fails more often than the ac unit from environmental contamination.

The most troublesome aspect of servicing the ac drive control was due to its location on the machine, said 35% of the respondents. And unavailability of published service data was cited as the biggest problem in servicing both ac and dc units, said 25%. The need for complex or expensive inventories of spare parts caused the greatest service problem for 33% of ac drive control users and 36% of dc drive control users. Repairing an ac unit was hampered by unavailability of spare parts, said 25%, and 32% felt that their major difficulty in getting ac drive control up again was the excessively high proficiency required of service technicians. Percentages add up to more than 100% because of multiple answers.

Rotary limit switches control sequential operations

Solid-state cam actuated rotary limit switches, by Emerson Electric Co.'s Commercial Cam Div., have sensing elements that utilize photoelectric interrupter modules to provide long life and high-speed switching without contact bounce. CT 4000 Series features 4, 8, and 12 switch units and dc sinking outputs, with 12 to 30-vdc input and 200 ma output per switch. CT 6000 Series has 4, 8, and 12 switch units, plug-in ac and dc output modules that can be mixed in any ratio, and 115 vac input. Both series have precise switching capabilities, LED

indicators to show output is "on," and degree wheel and pointer for easy setup and adjustment.

tion. Many mentioned the need for troubleshooting guides, others cited illustrations and flow charts for helping to track down problems, and several pointed to the need for self-diagnostics. Several readers also suggested that more control component manufacturers should provide guidelines for establishing preventive maintenance programs.

Tied for first place as most-mentioned, also by 22% of the respondents, was the suggestion that control component manufacturers do more testing and quality control to help prevent failure-prone units from leaving their plants. And another 14% felt that lead times for obtaining spare parts are often excessive; several mentioned lead times of several

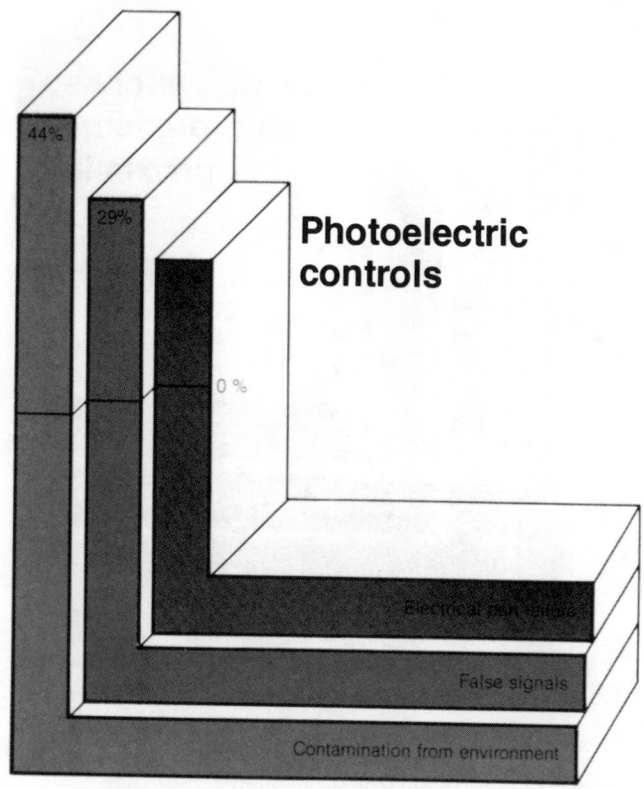

Photoelectric controls

44%
29%
0 %

Electrical part failure

False signals

Contamination from environment

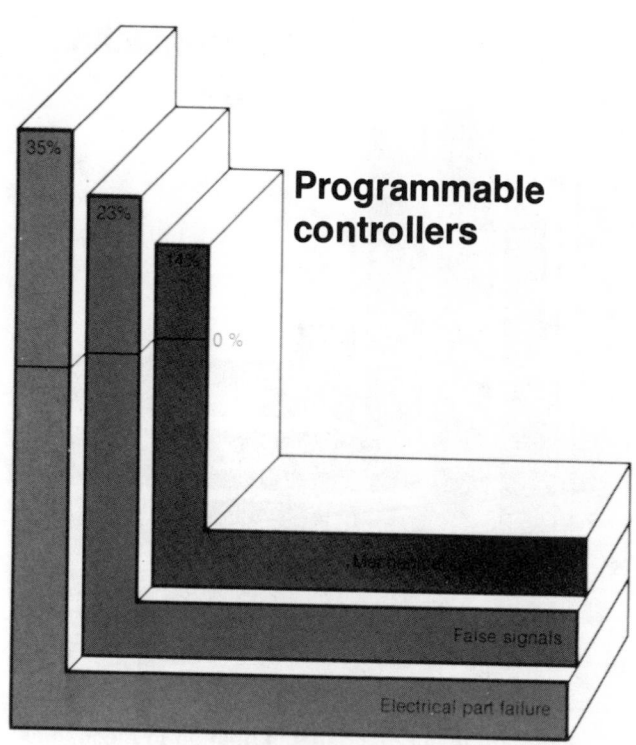

Programmable controllers

35%
23%
14%
0 %

False signals

Electrical part failure

Environmental contamination does a photoelectric control in more often than anything else, according to 44% of the respondents. False signals are the most common culprit in 29% of the respondents' plants, while another 19% cite failure of electrical parts. Programmable controllers, on the other hand, are most often done in by electrical part failures, said 35% of the respondents, with false signals the predominant failure mode for another 23%. Surprisingly, the third most common PC failure mode was loosening of mechanical parts. The need

for an expensive or complex spare parts inventory and the location on the machine were cited as being the most troublesome aspects of restoring failed photoelectric controls to service by 17% and 19% of the respondents respectively. And, while 37% said the worst problem in servicing PCs was the excessively high proficiency required of service technicians, only 9% said that published service data are unavailable. The need for a complex and expensive spare parts inventory was also picked by 23% as their biggest problem with PC service.

weeks and more for critical parts. Controls manufacturers could also improve the availability of their service personnel, according to 7% of the respondents, some of whom indicated that even manufacturer's representatives were not able to correct problems expeditiously.

Controls that are simpler, and easier for people with lower skill levels to maintain, were called for by 14% of the respondents, and another 10% pointed out that standardization of parts, lower parts counts, and increased use of modular parts would help their downtime-reduction efforts a great deal.

And even allowing for the natural human tendency to ask for perfection and Utopia, comments on the need for better quality, greater

Microcomputer industrial timer has on-board diagnostic program

On-board diagnostic program simplifies maintenance of industrial microcomputer-based timer. The built-in self-test program permits the user to verify proper operation of the Series 365 Long-Ranger, by Automatic Timing & Controls Co., without test instruments. Instructions are printed on a self-contained flip-up card so the user can determine if a problem is internal to the timer or in external circuits or relays. Unit is totally adjustable between 0.01 sec and 999 h, with a repeat accuracy of ±10 ms at all settings. It can be programmed to time up or

down from the set point, and can be adjusted during a timing cycle. Internal microcomputer also provides high immunity to electrical noise.

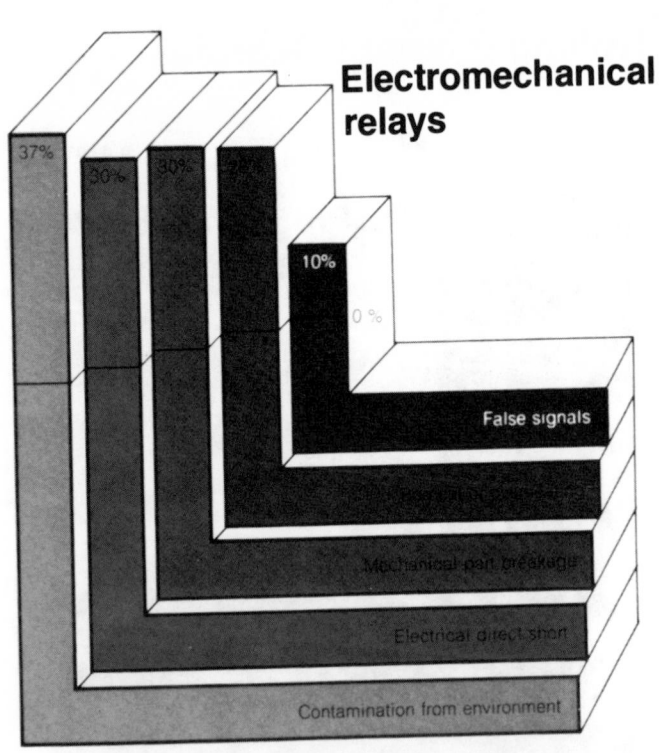

Electromechanical relays

37%
30% 30%
10%
0 %

False signals

Mechanical part breakage

Electrical direct short

Contamination from environment

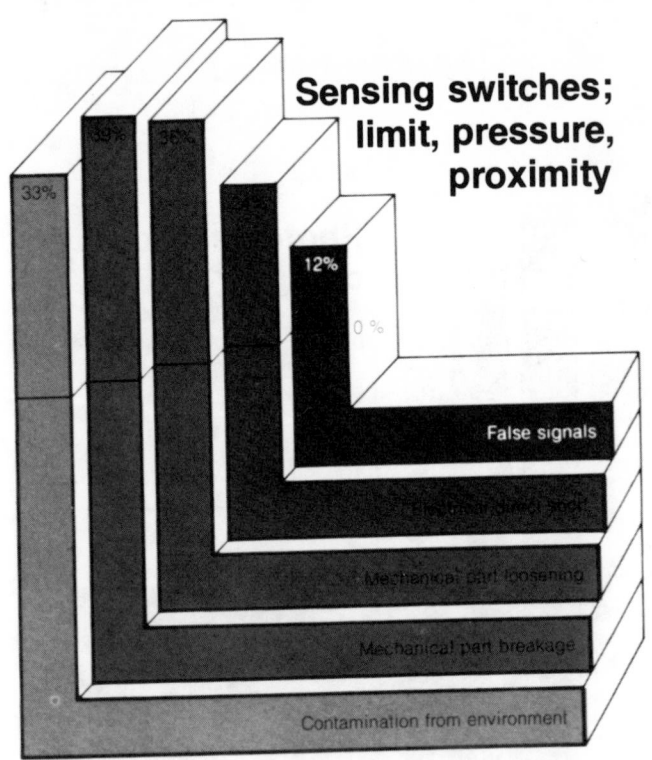

Sensing switches; limit, pressure, proximity

33%
39% 36%
12%
0 %

False signals

Mechanical part loosening

Mechanical part breakage

Contamination from environment

Although many respondents pointed to mechanical problems as being the most common failure modes in these predominantly mechanical devices, which is to be expected, 37% said environmental contamination is the most common failure mode for electromechanical relays. And 33% picked it as the most common sensing switch failure mode. False signals were the least common failure modes for both groups, again to be expected. Electromechanical relays appear to fail about equally often from mechanical breakage, burnout or overheating, and electrical direct short, while sensing switches seem more vulnerable to

mechanical than to electrical problems.

The most troublesome aspect of dealing with failed relays was, again, the expense and complexity of spare parts inventories, according to 15% of the respondents. And 20% pointed to the location on the machine as the biggest problem in servicing the sensing switches, which one might expect from the nature of the devices and the ways in which they are used. Other service difficulties drew little comment and over 20% of the respondents said they have no service problems with either of these categories of control components.

reliability, and longer mean-times-between-failures voiced by 21% of the respondents raise the question of whether control component and system reliability are all that the state-of-the-art would allow.

Buy the best

And, finally, readers were asked how their own company managements could better support their downtime-reduction programs. Buy the best, was the consensus of 12% of the respondents, who suggested that looking beyond the lowest bidder is often the most economical in the long term, and that investing in first-class equipment is actually an invest-

ment in downtime prevention.

Much of the response suggested that management gives preventive maintenance short shrift too much of the time. Tradeoffs are always made between today's production and tomorrow's problems. And it seems that management sometimes expects high uptime figures while simultaneously limiting the anti-downtime budget, because 43% of the respondents called for more management support for preventive maintenance programs and for management approval of capital allocations to permit stocking adequate supplies of spare parts.

Another 6% of the respondents specifically called attention to the need to reduce downtime by replacing aging and obsolete control components and systems. Others

suggested that computerized downtime reports and parts usage profiles would also be helpful.

On the other side of the coin, a number of respondents indicated that they have almost *carte blanche* in their downtime-reduction campaigns. One stated that his company would endorse any effort his group proposed to reduce downtime, while others commented that preventive maintenance and minimum downtime are already universally emphasized throughout their companies.

Clearly, the downtime problem will not go away overnight if, indeed, ever. But knowledge is power, and knowledge about control component and system failures represents the power to prevent them, or at least minimize their adverse effects.

Reprinted courtesy of Charles Berggren of Pennsacola, Florida

Malfunction Diagnosis of Rolling Element Bearings

J. Charles Berggren
Monsanto Company
Pensacola, Florida 32575

ABSTRACT: Rolling element bearings usually go through several distinct phases
of degradation prior to failure. The first phase is characterized by high frequency ultrasonic noise, followed by audible noise, then significantly measurable vibration, and finally, excessive temperature. Diagnostic techniques
available for detecting bearing faults in each of these stages, features of
the instrumentation utilized, and the relative advantages, disadvantages, and
costs of each method are discussed. Specific examples of faulty ball bearings
detected with an acoustic spike energy instrument are described in detail.

Key Words: Acoustic spike energy; diagnostic techniques; instrumentation;
noise; rolling element bearings; ultrasonic; vibration

Introduction: Rolling element bearings are used in virtually every piece of
equipment that a plant engineer is responsible for maintaining. Each of these
bearings has a rated B-10 life by the bearing manufacturer, which is rarely
attained in actual service. Excessive vibration, lack of proper lubrication,
external contamination, and faulty installation methods all enter into the
picture for reducing a bearing's life. As the condition of a bearing deteriorates because of these reasons and other factors, it gives off warning signals
prior to failure. This paper describes the distinct phases of degradation most
bearings experience prior to failure, means for diagnosing the severity of the
fault, advantages and disadvantages of these diagnostic methods, and relative
costs involved. Using this information as a guideline, the plant engineer can
decide which methods are most appropriate for his specific equipment.

PHASES OF BEARING DEGRADATION

New bearings and used bearings in good condition usually exhibit little or
only moderate audible noise. If the vibration spectrum is viewed on a real
time analyzer, a broad band of low amplitude frequencies can be seen. Should
the bearing be listened to with a stethoscope or a metallic object pressed
against the ear, the sound will be similar to that of ocean surf.

When the bearing begins to develop mechanical faults, such as scratches on the
raceways, hairline cracks in the cage assembly, or microscopic spalling of the
balls or rollers, high frequency ultrasonic sound and vibration is emitted.
As these faults start to become more prominent, the bearing is literally
"screaming under stress." The ultrasonic sound and acoustic spike energy will
start to rise sharply. At this stage the human ear can sense no change in
sound, and vibration analyzers will detect no significant change in bearing
housing displacement or velocity. Only an instrument specifically designed
to pick up and measure these ultra high frequency spikes will detect these
faults from their outset. Figure 1a shows a typical bearing inner race with
defects in the initial development stage.

As the bearing faults become readily visible to the naked eye, an audible
change in the bearing noise can be discerned by the trained ear. An accelerometer transducer, used in conjunction with a real time analyzer (RTA), can

now detect these faults. Data must be viewed initially in the logarithmic mode to pick up the inner and outer race ball pass frequencies, ball spin frequency, or rollers and cage assembly. This is necessary because the vibration signature noise floor in the linear mode is usually nearly as high as the discrete frequency peaks of these faults. Figure 1b shows a bearing with these defects just becoming clearly visible.

After this point, audible noise from the bearing can be readily heard. If a screwdriver pressed to the ear or a stethoscope is used to listen to the bearing noise, the sound is very much like that of rocks tumbling down a metal chute. When this point of degradation is reached, the bearing frequencies can be readily picked up using a standard velocity transducer with a manual tunable filter analyzer or with an RTA in the linear mode. Figure 1c is typical of a bearing in this stage of damage development.

The length of time a bearing can last until failure once it reaches this stage is largely dependent on the machine speed, load, lubrication factors, and environmental conditions. Good predictive maintenance programs will detect a bearing fault prior to or shortly after onset of the audible noise phase. If the machine is critical to the process and cannot be shutdown promptly for repair, the defective bearing can be monitored at daily to weekly intervals to closely watch for any large step change in condition. When deterioration in the bearing condition can be measured over a 24 hour period, life expectancy is usually a matter of less than a week.

If the bearing is allowed to continue in operation, one of two patterns usually emerges. Often in the case of low speed bearings, as the defects become worse, the bearing will "wear in." High frequency ultrasonic sound, acoustic spike energy, and RMS vibration velocity will all diminish, but vibration displacement at the running speed will greatly increase. This is only the "calm before the storm", however, since the ever increasing roughness within the bearing will eventually cause it to heat up, bake out the lubricating oil or grease, and then suddenly seize up and fail. In high speed bearings, the increasing surface deformations caused by defects usually manifest themselves in steadily rising ultrasonic, vibration, and noise levels. Once the bearing starts to overheat, failure is often only hours and sometimes just minutes away. Figure 1d shows a bearing with gross damage that is subject to seizure at any time.

AVAILABLE DIAGNOSTIC METHODS

A wide variety of diagnostic methods can be utilized to monitor the condition of rolling element bearings. These methods range from simple changes in audible and ultrasonic sound, to complex vibration analysis measurements that can pinpoint the specific defective components in a bearing. This paper will address the following commonly used diagnostic techniques with which the author has personal experience, giving a brief description of the devices utilized, relative advantages and disadvantages of each method, and the approximate costs involved:

1. Change in audible sound
2. Temperature measurement
3. Vibration analysis
4. Change in ultrasonic sound
5. Shock pulse method
6. Spike energy method
7. Acoustic emission method

TYPICAL PROGRESSIVE STAGES OF BEARING DEGRADATION

A

INITIAL PHASE

Noise level normal
Temperature normal
Vibration normal
Noticeable increase in ultrasonic sound, acoustic emission, and spike energy

B

SECOND PHASE

Slight change in noise level
Temperature normal
Slight increase in vibration acceleration
Large increase in ultrasonic sound, acoustic emission, and spike energy

C

THIRD PHASE

Noise level quite audible
Slight increase in temperature
Large increase in vibration acceleration and velocity
Very high ultrasonic sound, acoustic emission, and spike energy levels

D

Fig. 1—a,b,c,d. *Progressive stages of spalling caused by inadequate lubrication.*

FINAL PHASE

Change in pitch of noise level
Significant temperature increase
Significant increase in vibration displacement and velocity
Gradual decline followed by rapid increase of acoustic emission and spike energy

Other proven techniques, such as the crest factor, stress wave energy, and optic fiber methods, are also available. However, the author has no direct experience with this instrumentation and therefore will not comment on their effectiveness. The vibration analysis and spike energy methods have proven to be especially effective in our nylon chemical intermediates and yarn plant, and are discussed in the greatest detail.

CHANGE IN SOUND METHOD

The simplest and oldest method of detecting faulty bearings is listening to the change in audible sound the bearing produces. This is commonly done using a screwdriver, with the point pressed firmly against the bearing housing and the end of the handle held against the ear. A more sophisticated method employing the same principle is the industrial stethoscope. This low cost instrument (under $300) is usually battery powered to amplify the sound.

Advantages of this method are its simplicity and minimum investment. There are many disadvantages, however, that limit its usefulness. First, change in sound is a highly subjective diagnostic tool. Different mechanics will interpret the noises they listen to in a different manner. Cavitation, aerodynamic turbulence, and other noises produced by the machine being inspected often interfere with the noise emitted from the bearing. This is especially true when there is a significant distance between the screwdriver or stethoscope tip and the bearing. Secondly, trend analysis is inaccurate because no meter readings or hard copy recordings are made to chart the change from one inspection to the next. A deteriorating bearing condition will certainly be picked up by an experienced mechanic or technician using a stethoscope, but the relative severity of the change cannot be calculated. Thirdly, detection of a bad bearing is often too late in the case of a critical machine which must remain in operation until the next planned turnaround. Bearing damage is already well advanced, and Maintenance is left with little planning time.

In practice, the author has found the change in sound diagnostic method to be occasionally useful for physical confirmation of a meter reading. The technique is most effective when utilized by an experienced mechanic or technician with a "trained ear." Progressive plants with critical equipment need much more sophisticated means, however, to accurately pinpoint bad bearings.

CHANGE IN TEMPERATURE METHOD

A second commonly accepted diagnostic method is measuring the change in temperature of a bearing. This is done by feeling the bearing cap with your hand, using a portable pyrometer or digital temperature meter, or reading the output of permanently installed thermocouple monitors. Figure 2 shows some typical meters commercially available.

This method is easy to use and interpret. Temperature is a well accepted condition indicator since excessive heat usually means a serious fault exists. Portable temperature meters are also inexpensive ($200 to $400 including contact probe).

Fig. 2

There are two major disadvantages of using temperature as the primary device for detecting bad bearings. First, fault detection is often too late since once a bearing starts to overheat, its future life is very short. This gives the maintenance department insufficient reaction and planning time. Second, trend analysis is too variable since bearing temperature will change with machine load and ambient temperature. Temperature measurement is most valuable for detecting an assembly fault after a new bearing is installed on a machine, or after a significant speed increase or load change is made. High speed spherical roller and tapered roller bearings are particularly sensitive to improper mounting, and excessive operating temperature shortly after startup is a sure sign of trouble. Temperature measuring instruments are recommended as a supplementary tool for plant predictive maintenance programs.

VIBRATION ANALYSIS METHOD

Using vibration analysis to detect faulty bearings is a well accepted and widely used diagnostic method. Instruments used include portable vibration meters (Figure 3), vibration analyzers with manually tunable filters (Figure 4), and real time spectrum analyzers (RTA), (Figure 5). These devices employ either a velocity pickup or accelerometer as the transducer, and provide the analyst with a conventional vibration readout in inches per second velocity or g's acceleration. The reading is usually displayed on an analog meter, although more recently developed models can provide a digital readout.

Inexpensive, handheld vibration meters that display only the overall unfiltered vibration are limited in their effectiveness to accurately diagnose bad bearings. Vibration caused by unbalance, misalignment, looseness, worn belts, and other sources mask the signals being emitted by a failing bearing on a machine. For this reason, the vibration signal must be filtered and the frequency spectrum analyzed in order to pick up the bearing frequencies. A vibration analyzer or RTA is required to separate the vibration signal into discrete frequencies for detailed analysis.

The vibration analysis method provides very accurate trend analysis. The type and location of the fault can be pinpointed, and its relative severity reliably assessed. The technology has a proven track record, and this method can be counted on to produce consistent results. James Taylor has published several excellent papers that enable a skilled technician or engineer to even predict

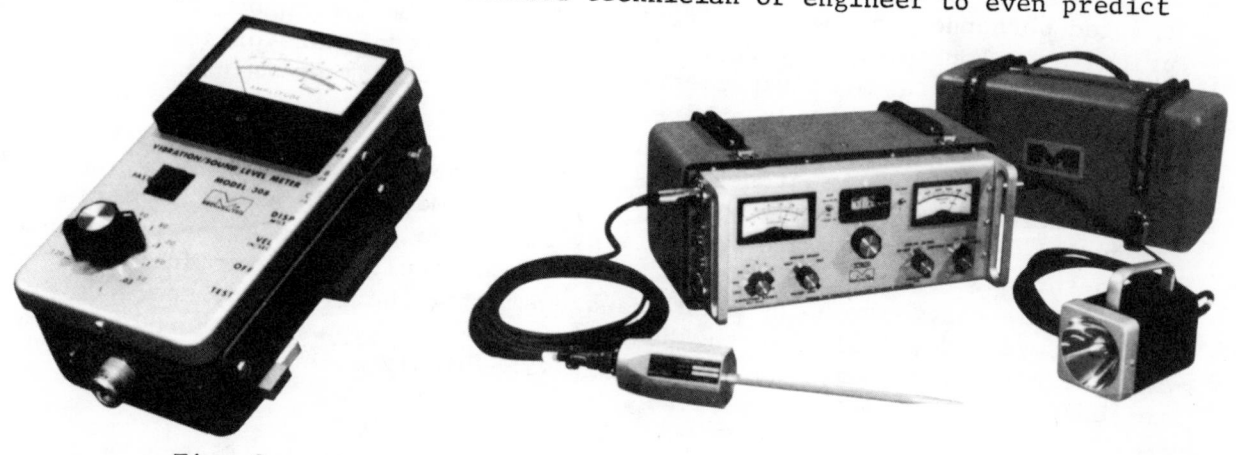

Fig. 3   Fig. 4

whether the bearing defect is located in the raceways, rolling elements, or cage. Several leading vibration analyzer manufacturers have published very good troubleshooting charts to aid the vibration analyst.

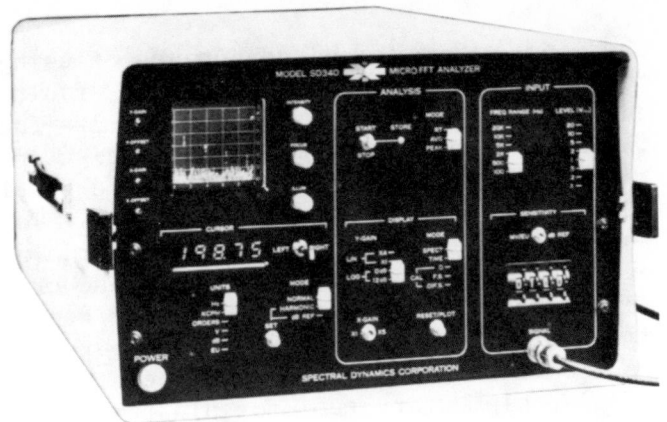

Fig. 5

The vibration analysis method has a number of drawbacks that impair its effectiveness, however. First, fault analysis is very time consuming, especially with a vibration analyzer using a manually tuned filter. Secondly, instrumentation usually weighs 20 to 50 pounds. Though portable, transporting instruments this size quickly becomes very tiresome. This problem can be alleviated somewhat by taping the data with an instrumentation quality FM recorder (Figure 6), and playing the tape back into an RTA in the office. Even these tape recorders can become a burden, however, when the analyst must climb up and down many stairs in making his rounds. Third, vibration analysis instruments are quite expensive.

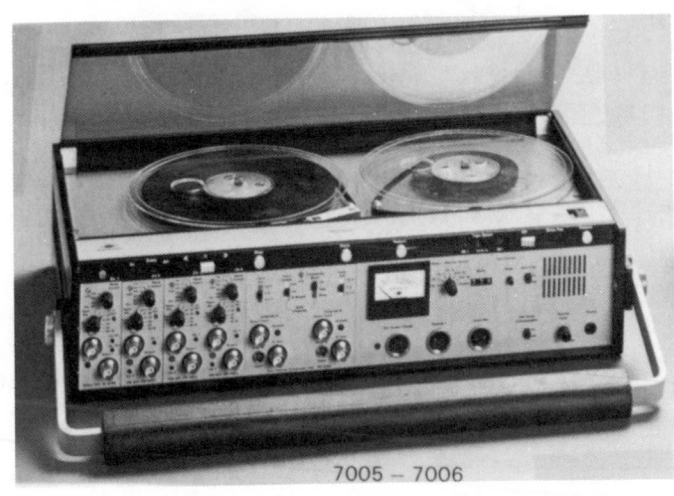

7005 - 7006

Fig. 6

Manually tuned analyzers start at $3,000. Sophisticated real time analyzers cost from $6,000 to $25,000. This is not an obstacle for large plants with extremely valuable rotating equipment, but many small facilities, without skilled personnel tained in vibration analysis, simply cannot afford this price tag. Fourth, highly trained maintenance technicians or mechanical engineers with practical experience in rotating equipment maintenance are needed to operate vibration analysis instruments and interpret the data to achieve good consistent results. Again, this can be a serious problem for small plants. Fifth, most bearings will not show significant vibration until well after the fault has developed. By this time, the bearing has increased in audible noise level, and may already have been reported by area mechanics or operators. Vibration analysis confirms the fault and determines its severity in this case. By using an RTA and examining the frequency spectrum in the logarithmic mode, faults at the bearing frequencies can be detected earlier. This requires more time, however, along with highly experienced and alert personnel to perform the analysis.

The author has extensive first hand experience using many different models of handheld vibration meters, analyzers with manually tuned filters, and real time spectrum analyzers. Based on this experience, vibration analysis has its greatest value when used to obtain more detailed fault data originally found by a more portable instrument designed exclusively for measuring bearing defects. Real time analyzers are far superior to manually tuned filters since the whole frequency spectrum can be viewed at one time, widely fluctuating amplitudes can be easily averaged, peak values can be accurately determined, and the waveform can be readily analyzed on the built in CRT screen. Accelerometers are preferred to velocity transducers because of their wider frequency range and smaller size. Hand holding the transducer or using a magnetic mount is usually satisfactory for routine screening inspections. Mounting the transducer to the bearing cap with dental cement or a stud against a flat clean surface is necessary, however, to obtain accurate repeatable data for critical equipment, particularly very high frequency data over 2,000 hertz. In summary, vibration analysis is most effective for detailed follow-up analysis of a defect originally found by an incipient failure detection instrument.

ULTRASONIC MEASUREMENT METHOD

Portable ultrasonic meters (Figure 7), are a fourth device for detecting antifriction bearing faults. These lightweight, handheld instruments employ a contact probe with a piezo-electric crystal pickup that resonates in the presence of ultrasonic signals in the 36 to 44 kilohertz frequency range. The signal output is displayed on a small 0-100% range analog meter with a manually adjustable amplifier gain dial. A set of earphones is provided for the operator to listen to the signal as well. Impact noises from damaged rolling elements or rough raceways can be readily heard through the earphones. The meter dial will peg out at full scale at a lower gain setting when a bad bearing is compared to one in good condition.

Instruments of this type are very lightweight, easy to use, and do not require highly skilled personnel to operate. Cost is under $1,000, including standard accessories.

Several key factors limit their usefulness as a bearing condition indicator, however. First, the headphone sound and meter readout is dependent on the amplifier gain setting. Slight changes in the gain setting make a significant difference in the meter reading. Second, no industry established severity standards exist for ultrasonic readings as they do for vibration measurements. This makes it difficult to correlate the meter reading with familiar standards for comparison. Third, the ultrasonic meters do not provide a vibration readout in familiar displacement, velocity, or acceleration units.

The author has found ultrasonic meters to be the most useful for detecting air and gas leaks in piping systems, defective valves in reciprocating compressors, and for evaluating the condition of steam traps. Ultrasonic meters are superior to conventional stethoscopes for listening to bearings, but because of their drawbacks are not recommended as the primary diagnostic instruments for bearings.

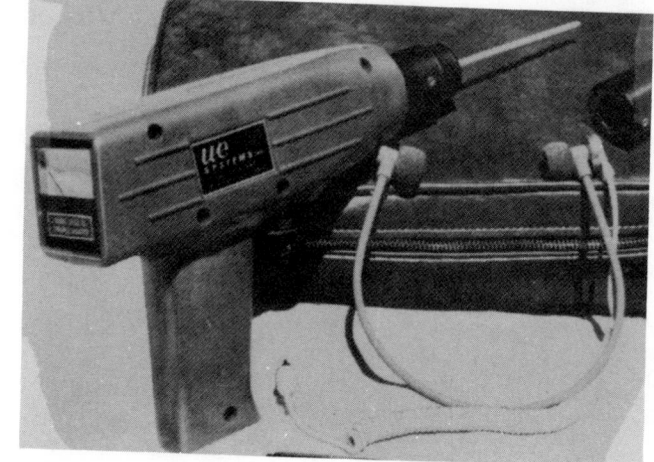

Fig. 7

SHOCK PULSE MEASUREMENT

The shock pulse method is a technique developed in the 1970's specifically for evaluating rolling element bearing condition. This diagnostic method is based on the principle that irregularities in the bearing surfaces cause shock pulses (extremely short duration pressure waves produced from mechanical impact) in the bearing and its housing during operation. The magnitude of these impact forces correlates directly to the bearing condition, and is not influenced by machine vibration induced by unbalance, misalignment, or hydraulic turbulence. Magnitude of these shock pulses is directly related to the roughness of the bearing running surfaces, as well as its size and rotating speed. To account for the bearing geometry and speed, the bore and RPM is dialed onto an externally mounted meter scale, (Figure 8), to obtain an "initialized value" expressed in dB. Then the additional shock pulse energy above the baseline level is determined by adjusting the meter dial until the continuous tones emitted by the instrument become intermittent. This dB reading, defined as the noise "carpet" level for the bearing, can be compared to the RMS reading of a vibration signal. As the dial is turned further, the tones eventually disappear. This maximum value is expressed in dB and may be compared to the peak reading of a vibration signal with numerous spikes on the waveform.

This method has many advantages for detecting faulty bearings. First, the bearing's condition can be accurately determined from established standards based on field experience. Trend analyses can be made to select the optimum replacement point. Second, the shock pulse method is an incipient failure detection device (IFD) that diagnoses faults early in their ultrasonic development stage. This allows ample time for maintenance planning. Third, the instrument is lightweight and easy to handle. Fourth, it serves as a good screening device, since the bearing condition can be determined in less than one minute. Fifth, cost is moderate (approximately $2,000).

The shock pulse method does have several disadvantages that limit its effectiveness. First, the bearing bore and RPM must be known to dial in the reference baseline dB level. Otherwise, inaccurate results will be obtained. This can be a significant problem if a large variety of equipment is being monitored, unless accurate machine data is obtained and listed in a ready reference form. Secondly, skilled mechanics or technicians are needed to operate the instrument to achieve consistently good results. Readings

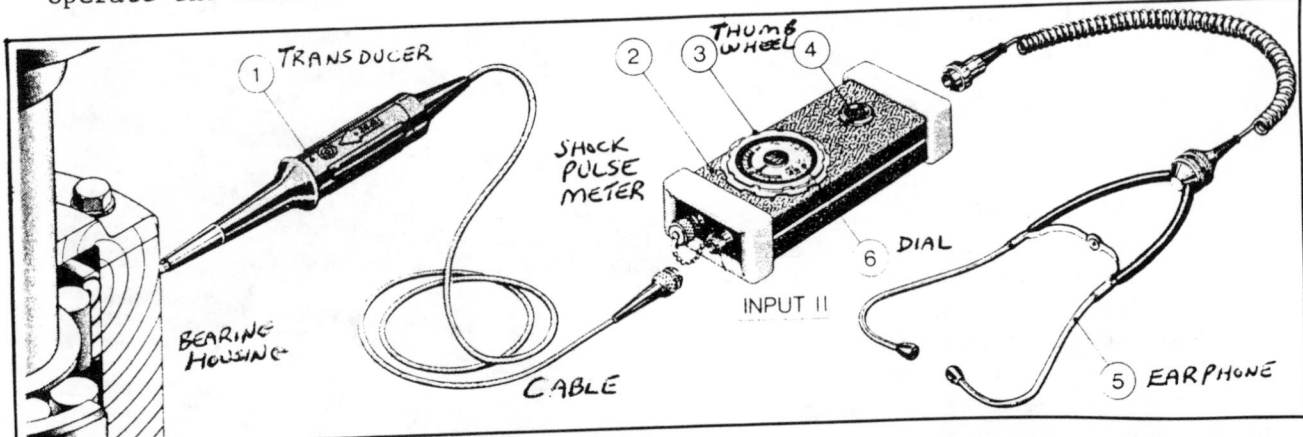

Fig. 8

obtained are very sensitive to transducer mounting and location. Thirdly, the instrument does not provide a conventional vibration readout. A second instrument is required for this purpose.

Several plants in the author's company report very good success with the shock pulse meter. It has a proven track record for detecting bad bearings. Time required to obtain the shock value, and the need to know the bearing size and speed, limit its effectiveness for rapid fault diagnosis.

SPIKE ENERGY METHOD

The spike energy method is another excellent technique developed in the late 1970's for detecting faults not only in rolling element bearings, but damaged gears and rotor rubs as well. Instruments classified as spike energy meters utilize a piezoelectric crystal accelerometer that resonates in the 25 to 35 kilohertz frequency range. A high pass filter is employed in the circuitry to reject those frequencies below the natural frequency of the accelerometer. This accentuates the ultrasonic frequencies that are most indicative of kinetic and friction energy that is generated by defects that cause mechanical impacts in the bearing assembly. An analog meter, (Figure 9), displays the spike energy measured in units of G's acceleration or decibels. An oscilloscope or strip chart recorder may be connected to an output jack on the instrument for more detailed analysis of the signal. Another model of the instrument, (Figure 10), can produce a strip chart record of the spike energy or vibration measurements.

The spike energy method has many advantages for early detection of faulty rolling element bearings. First, these instruments are incipient failure detection devices that can sense and measure a bearing defect long before noticeable vibration or audible noise develops. Second, trend analysis capability is very good. The rate of degradation can be accurately plotted and tracked to select the optimum replacement point in relation to planned equipment shutdown. Third, these instruments make excellent screening devices for routine inspections in a predictive maintenance program since meaningful data indicative of the bearing condition can be quickly obtained. They have the added benefit of providing a conventional vibration readout in units of inches per second velocity.

Fig. 9

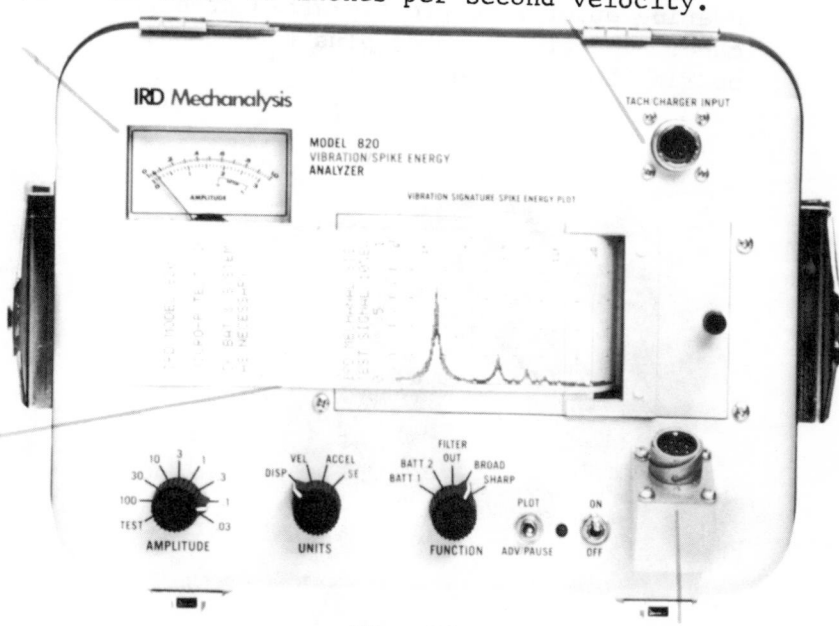

Fig. 10

Fourth, the meters weigh less than 5 pounds. This light weight and portability make them very easy to transport. They do not burden down the analyst when he has to make several hundred measurements over the course of a day. Fifth, meter cost is modest, less than $2,500. Sixth, mechanics and technicians can be quickly trained to obtain reliable, consistent data.

There are some negative aspects of this method, however. First, the readout is sensitive to the measurement location of the transducer. The same point must be utilized for each subsequent inspection to achieve repeatable data. Second, the method is sensitive to other ultrasonic signal generating sources such as cavitation in pumps, control valve noise, insufficient bearing lubrication, and rubbing of shafts and sheaves against a guard that will also produce a high spike energy signal.

The author has found these instruments to be highly effective for routine monitoring of rotating equipment, especially motors, fans, pumps, centrifuges, gearboxes, and textile spinning equipment bearings. Care must be exercised, however, to distinguish cavitation noises from bearing faults on pumps, and to be alert for external disturbances such as guards rubbing or belt squeal. Figures 11-14 show four typical defective bearings that were detected using the spike energy method.

Figure #11 shows parts from a 6311 deep groove ball bearing taken from a 50 HP, 1750 RPM, centrifugal pump in fire water service. The bearing had shown a sudden jump in vibration to .45 in/sec velocity and .34 G's spike energy. Both readings were quite erratic. The inner and outer bearing races were badly spalled. One of the balls had literally cracked apart, and the others were visibly etched.

Figure #12 shows a 6315 C3 deep groove ball bearing taken from a 300 HP, 1180 RPM, vertical river water centrifugal pump. The vibration reading taken near the top of the motor housing was only .27 in/sec velocity, but the spike energy value had jumped from .06 to .20 G's in only two weeks. The technician also noticed a distinct change in sound coming from the motor. The inner race shows heavy spalling, and several of the balls have deep gouges in them. The long mechanical transmission path from the outside of the motor housing to the bearing outer race greatly attenuated the overall spike energy reading in this case. The step change, plus the technician's experience, enabled the defective bearing to be found.

Fig. 11

Fig. 12

Figure #13 shows a 6309 deep groove sealed ball bearing removed from a 25 HP, 1750 RPM, electric motor. The motor bearing indicated a sudden jump to 1.30 G's spike energy and an overall vibration velocity of .40 in/sec. When pulled from the motor, inspection revealed the bearing had been turning in the end bell housing.

Fig. 13   Fig. 14

Figure #14 shows a SAF 22318 pillow block bearing housing removed from a 75 HP, 1780 RPM, belt driven exhaust blower. The blower bearing showed an increase in vibration velocity to 1.10 in/sec and a spike energy of .83 G's. Upon removal for inspection, the spherical roller bearing was found to have been rotating in the pillow block housing with .006" total diametrical clearance. Replacement of the pillow block and bearing lowered the vibration to .15 in/ sec velocity and the spike energy to .05 G's.

Figure #15 is a vibration and acoustic spike energy tolerance chart that was developed for use at our plant. The relative severity of vibration velocity and acoustic spike energy readings are classified to assist Manufacturing and Maintenance personnel in their interpretation. Values shown are for typical ball and roller bearings where the transducer can be mounted directly on top of the bearing housing. Where the bearings are deep within the machine casing, the values must be factored down somewhat to account for mechanical transmission losses in the signal.

ACOUSTIC EMISSION METHOD

The acoustic emission method is still another excellent technique similar in concept to the spike energy method. Acoustic emission detectors, (Fig. 16), acoustic flaw analyzers, (Figure 17), and other instruments classified as incipient failure detectors also utilize a high frequency piezoelectric crystal transducer. However, these instruments employ a 80-120 kilohertz band pass filter in the circuitry to detect those acoustic emission frequencies that experience has found to be very responsive to deformations in metals subjected to stress in the bearing. An analog meter displays the acoustic emission energy in units of root mean square (RMS) amplitude and signal above threshold

FIG. 15

MONSANTO COMPANY, PENSACOLA, FLORIDA
MAINTENANCE DIAGNOSTICS VIBRATION AND ACOUSTICS GENERAL TOLERANCE CHART FOR PROCESS
EQUIPMENT WITH ROLLING ELEMENT BEARINGS

DIAGNOSTIC SUPERVISOR 7565

DIAGNOSTIC TECHNICIAN 7419

VIBRATION VELOCITY INCHES/SECOND	ACOUSTIC SPIKE ENERGY		SEVERITY	INTERPRETATION	CORRECTIVE ACTION REQUIRED
	DB	G's SE			
1.5+	50+	2.0+	Danger	Very critical fault, equipment unsafe to operate, failure imminent, immediate shutdown required to prevent failure in service.	Schedule mandatory shutdown within 24 hours with Operations. Expedite procurement of needed spare parts to reduce downtime.
.75 - 1.49	40-49	1.00 - 1.99	Very Rough (ALERT)	Serious fault, rapid wear over ensuing weeks, failure probable within 30 days if unit remains in operation, plan early shutdown for correction.	Contact Maintenance Diagnostics technician for detailed analysis. Schedule early shutdown with Operations. Expedite procurement of needed spare parts.
.40 - .74	30-39	.50-.99	Rough (ALERT)	Major fault, considerable wear over ensuing months, failure possible, schedule correction at next planned shutdown.	Contact Maintenance Diagnostics vibration technician for further analysis. Monitor trends weekly. Procure necessary spare parts if required.
.20 - .39	20-29	.20-.49	Fair (ACCEPTABLE)	Minor fault, gradual wear over long term, failure unlikely, correction not economical at this time.	None
.10 - .19	10-19	.10-.19	GOOD (ACCEPTABLE)	Insignificant fault, very little wear over long term, typical of equipment in normal operation.	None
.01 - .09	0-9	.01-.09	Smooth (ACCEPTABLE)	No fault, well balanced and well aligned equipment, typical of new installation.	None

(SAT). The RMS reading provides a relative measure of the average acoustic emission energy. The SAT reading is a relative measure of only the very intense bursts of acoustic energy. A sharp increase in this value is a sure sign of a developing defect in the bearing.

The acoustic emission method is perhaps the most sensitive of all the bearing fault detection techniques. A bearing flaw of any type will immediately show up with this instrumentation. Defects will be signalled before any measurable increase in vibration, audible change in sound, or excessive rise in temperature. Trend analysis is very accurate. The SAT parameter signals initial development of a defect. An increase in relative bearing roughness (RMS) parallels increases in measurable vibration and noise level. These instruments also provide a conventional vibration velocity readout, which is very useful. They are quite portable, weighing about 4 to 18 lbs. With a shoulder strap, they make good devices for periodic field measurements. Cost is modest, ranging from about $1,500 to $3,500. Heinz Bloch engineered an extremely sophisticated computerized system based on this acoustic emission detection principle, and reports excellent diagnostic results.

Like other methods, acoustic emission detection has its drawbacks. First is its extreme sensitivity. Since faults are detected so quickly, many bearings with considerable life remaining may be pulled. Up to date records must be maintained to determine the optimum replacement point at which the bearing is about to fail. Second, the extremely high frequencies being measured necessitate rigid, uniform measuring locations for the acoustic pickup. Spot-facing every bearing to be monitored and attaching a threaded stud can be very expensive if a very large number of bearings are to be monitored. Third, signal cables and transducers are quite fragile and cannot withstand rough treatment. Fourth, skilled personnel are needed to obtain consistent measurements and analyze the date for optimum results.

The author has found this technique to be one of the most sensitive of all methods for incipient failure detection. However, because of the stringent mounting requirements to obtain repeatable, meaningful data, this method appears most suited to permanent monitoring installations (Fig. 18) of critical equipment. The technique lends itself well to centralized computer control systems, and should be valuable for early failure detection of remotely installed machinery.

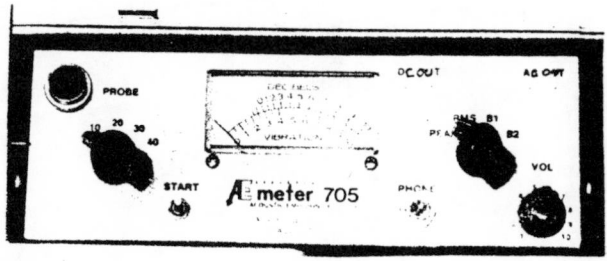

Fig. 16

Fig. 17

SENSOR & PREAMPLIFIER

MONITOR UNIT

Fig. 18

SUMMARY

From this discussion of available diagnostic methods, the plant engineer can
see that no one single technique is ideal for all equipment applications.
Each method has its pros and cons. The limitations of each method need to be
understood in order to achieve maximum results in a predictive maintenance
program. A combination of two or more of these methods is usually necessary
to provide sufficient diagnostic capability for all types of rolling element
bearing defects. Skilled, knowledgeable, and dedicated personnel to perform
the inspections, interpret the data, maintain accurate trend records, and
recommend corrective action needed will greatly improve results attainable.
Last, but not least, the support of Maintenance and Manufacturing supervision
is required to make the equipment available, and provide feedback on defects
actually found.

ACKNOWLEDGEMENTS

The author wishes to thank the engineering technicians in the Monsanto Maintenance Diagnostics Group at the Pensacola plant, especially Robert McGoun, Don Page, Tom Floyd, and William Griessel, who provided their invaluable assistance in data collection, assessment of equipment condition, and ideas for improving the efficiency of the alternative diagnostic methods discussed in this paper.

REFERENCES

1. Blake, Michael P., Vibration and Acoustic Measurement Handbook, Spartan Books, New York, 1972.

2. Taylor, James I., "Determination of Antifriction Bearing Condition by Spectral Analysis", The Vibration Institute sponsored Seminar on Machinery Vibration Monitoring and Analysis, February, 1978.

3. Taylor, James I., "An Update of Determination of Antifriction Bearing Condition by Spectral Analysis," The Vibration Institute sponsored Seminar on Machinery Vibration Analysis, April, 1981.

4. Barthel, Karl, "The Shock Pulse Method for Determining the Condition of Anti-Friction Bearings," The Vibration Institute sponsored Seminar on Machinery Vibration Monitoring and Analysis, April, 1979.

5. Parsons, Edward U., "The Role of Acoustic Emission as a Detection Technique," Allison Labs, Austin, Texas, 1975.

6. "Detection of Defects in Rolling Element Bearings and Gears," IRD Mechanalysis Application Report No. 114, IRD Mechanalysis, Columbus, Ohio, 1980.

7. Bloch, Heinz P., "Predict Problems with Acoustic Incipient Failure Detection Systems," Hydrocarbon Processing, October, 1977.

8. Finley, Robert W., "Incipient Failure Detection in Rotating Machinery," Chemical Engineering, July, 1980.

9. Taylor, Robin R., "Predict Bearing Failures with Portable Checkers," Hydrocarbon Processing, January, 1982.

10. Mitchell, John S., Machinery Analysis and Monitoring, Pennwell Publishing Company, Tulsa, Oklahoma, 1981.

11. SKF Industries Bearings Group, Bearing Failures and Their Causes, SKF Bulletin 148-110, March, 1980.

Reprinted courtesy of Bruel & Kjaer Instruments, Inc., Marlborough, Massachusetts

IMPROVED MACHINE MAINTENANCE
using
SIMPLE VIBRATION MEASUREMENTS

It is well established that changes in the vibration level of a machine indicate changes in the condition of the machine. Vibration levels will increase as the machine balance grade deteriorates, parts work loose, and bearing wear increases. The Integrating Vibration Meter Type 2513 is an accurate and economical instrument for making these day-to-day measurements. The easy to read display gives an indication of the overall vibration level as well as the "spikiness", thereby giving both condition, and additionally, diagnostic information.

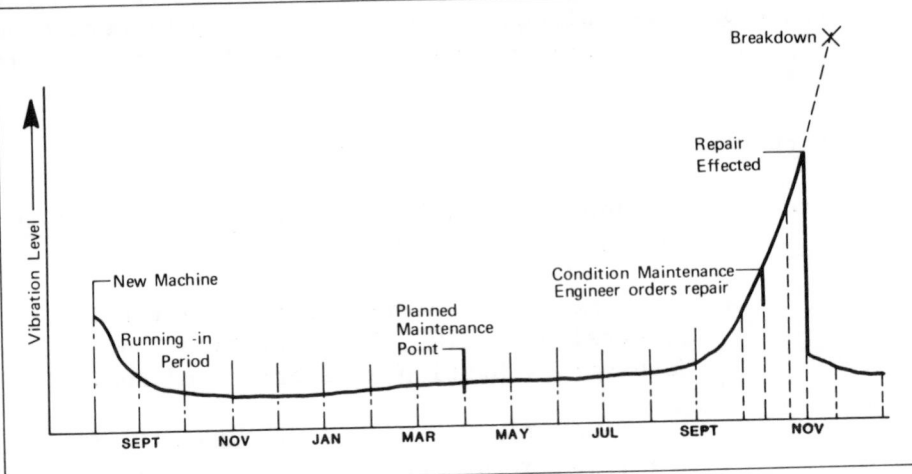

Figure 1. Vibration level *vs.* time for machine developing a fault.

Most mechanical faults in running machines are preceded among other things by an increase in vibration level. Regular vibration measurements will give early warning of deterioration and enable effective maintenance and overhaul to be planned before breakdown (Fig. 1). But there are several vibration parameters which may be selected for measurement. Each of these parameters tells us something different about the health of the machine. Which one or combination of these is best for monitoring over time is the subect of this application brief.

Figure 2 shows a typical vibrating signal from a rotating machine. Sharp spike-like peaks occur occasionally although seldom with the same height (amplitude). These peaks can be directly related to the max-

imum stress levels at the point of measurement. Another parameter which can be measured is the RMS level.* The RMS signal is directly related to the overall energy or power content of the vibration. Therefore, measuring RMS gives a better indication of overall vibration and machine condition than peak measurements. However, taken together, these two quantities (peak and RMS) give an even better characterization of the vibration signal than either one taken separately.

As an example, consider two machines of the same type. Both the peak and RMS vibration levels have been monitored regularly over time. One machine has a

small amount of mass imbalance on the rotor but is otherwise healthy. The second has developed pitted bearings and should soon be overhauled (Fig. 3). In both cases the RMS vibration levels will increase over time. The first machine because of the imbalance and the second because of the bearing wear. Examining both the RMS and Peak levels of the two machines will reveal something additional.

In the first case the vibration signal is dominated by the once-per-revolution imbalance component producing a vibration signal with a low ratio of peak to RMS. On the other hand, pitting in the bearing race causes many sharp impacts with the rolling elements resulting in a very impulsive signal with many large spikes. This signal is characterized by a much larger peak to RMS ratio than in the first case. By simply monitoring both the RMS and peak level periodically one can obtain more information as to the condition of the machine. For complete machine fault diagnosis a detailed frequency analysis of the vibration spectrum should be performed.

---

* The RMS or Root Mean Square level involves squaring, finding the mean value, and taking the square root of the incoming signal. Hence the resultant is termed the "root mean square"

$$\text{RMS Level} = \sqrt{\frac{1}{T}\int_0^T x^2(t)\,dt}$$

T = Averaging Time
x(t) = Instantaneous Vibration Level

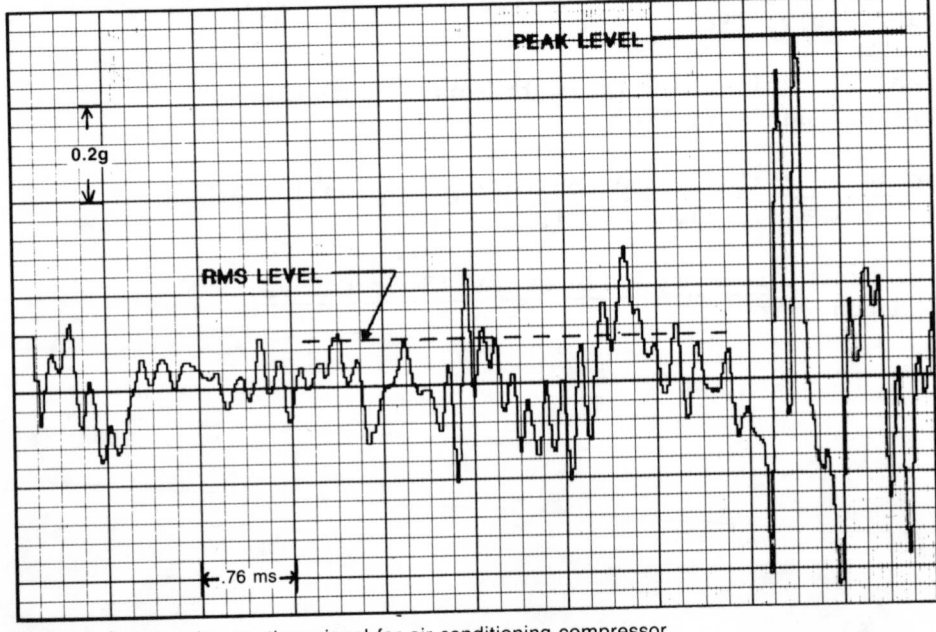

Figure 2. Acceleration *vs.* time signal for air conditioning compressor.

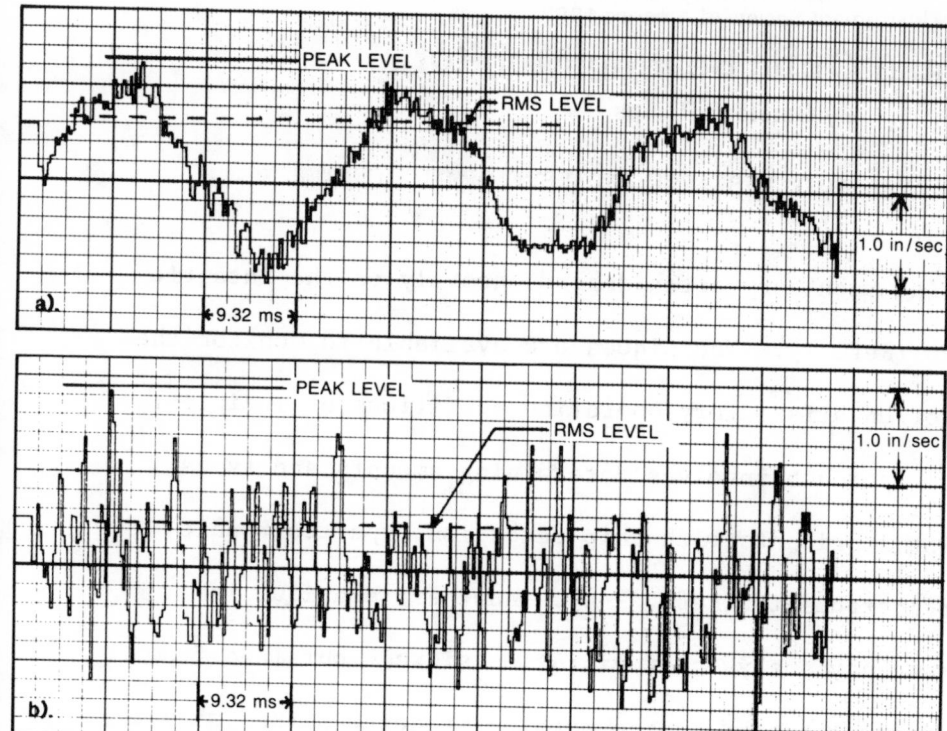

Figure 3. Vibration velocity *vs.* time signal for
a) Machine with a mass unbalance
b) Machine with bearing noise

The vibration level of a machine will often fluctuate so rapidly that finding a repeatable single number indication is difficult. In this case visual averaging of the fluctuating levels has been shown to depend significantly on the person reading the meter. The Integrating Vibration Meter Type 2513 solves this problem in its L_{eq} mode. L_{eq}† is the constant vibration level which has the same energy as the fluctuating vibration level. The reading built-up over a 60 sec. time period gives a true average energy level at the end of the period. The maximum peak value occurring over this time is simultaneously dis-

played. Since the L_{eq} is calculated over a 60 sec. time period, the short term fluctuations in the vibration signal are incorporated in the resulting value. During the 60 sec. period the slowly rising LED of the 2513 display blinks indicating that the L_{eq} calculation is still taking place. At the end of calculation the display stops blinking and remains on constantly for 1 minute to allow sufficient time to write down the result. The 2513 can be reset at any time to begin a new L_{eq} calculation. Therefore, repeatable measurements can be easily made even by non-technical personnel. The decision as to when the "L_{eq}" or "1 second RMS"

mode should be used may be made on a case to case basis. In general, the L_{eq} can always be used. The 1 second RMS position can be used when measuring relatively constant or slowly fluctuating vibration levels. If the signal fluctuations are too rapid to allow use of the RMS scale then the L_{eq} should be used.

The 2513 simultaneously indicated both peak and L_{eq} vibration levels in acceleration or velocity. These levels are easily read on its thermometer-like display. Therefore, with a single, simple measurement both peak and L_{eq} levels are obtained. Comparing these levels with previously recorded ones indicates the change in machine condition (Fig. 4).

Conducting vibration measurements with the 2513 is made even easier with the clipboard/holder supplied with the instrument. Space is provided for the meter and vibration pick-up as well as a pad of record sheets. The whole arrangement folds out for convenient side-by-side use of meter and recording pad. The Integrating Vibration Meter Type 2513 is a small lightweight instrument for making accurate day-to-day vibration measurements. Supplied with a precalibrated piezoelectric accelerometer, it can measure true-peak, maximum peak, RMS, RMS-maximum or L_{eq} in acceleration or velocity. In addition, both peak and either RMS or L_{eq} are displayed simultaneously with one simple measurement. Combining peak plus RMS, or L_{eq} has been shown to be advantageous when conducting a program of condition monitoring of vibrating machinery.

† The L_{eq} or equivalent continuous vibration level is obtained by electronically integrating the signal over a specific period of time.

$$L_{eq} = 10 \, Log_{10} \, \frac{1}{T} \int_0^T 10^{\frac{L}{10}} \, dt$$

T = Measurement Time Period
L = Instantaneous Vibration Level in Decibels.

The 60 second L_{eq} is equivalent to the RMS Level over a 60 second time period.

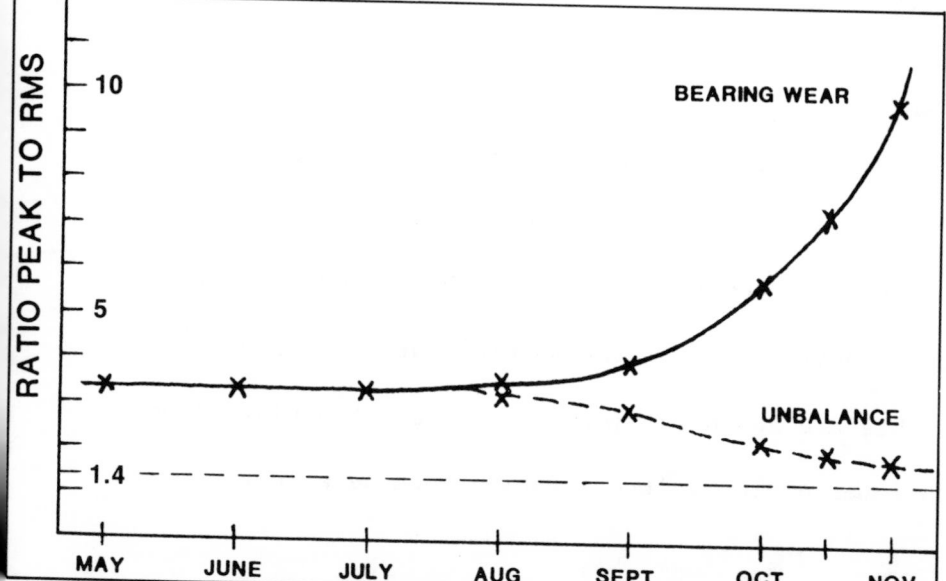

Figure 4. Ratio of peak to RMS *vs.* time for vibration signals from machine developing bearing wear and machine developing mass unbalance. (Overall vibration level as in Figure 1.)

Reprinted with permission.
From *InTech*, May 1982
© Instrument Society of America 1982

The Fiber Optic Bearing Monitor

By Gerald J. Philips

DAVID W. TAYLOR NAVAL SHIP RESEARCH AND DEVELOPMENT CENTER

Performance of rotating machinery often depends heavily on the integrity of ball or roller bearings (Ref. 1). Techniques are available to monitor the bearings for indications of unsatisfactory operating conditions or incipient failures. One traditional approach involves determining amounts and types of particles in the bearing lubricants with chip detection, spectrographic, ferrographic, and similar instruments (Refs. 2-4). Another technique is based on amplitude and frequency analyses of vibration signals obtained from accelerometers or other sensors mounted on the machine structures (Ref. 5-9).

FIBER OPTIC DISPLACEMENT SENSORS

Fiber optic displacement sensors offer an alternative way to monitor bearings (Refs. 10-15). Probes could be installed as in Figure 1, with one sensor focused at the stationary outer race through a hole in the housing and a second placed near an exposed area of the shaft. The first sensor detects displacements caused by the rolling elements in the bearing; the second indicates rotational speed and provides reference signals for ascertaining phase relationships.

Rotational Frequency Vibrations

The phase and amplitude of rotational frequency vibrations are often measured for in-place balancing. This is usually done with the rotor mounted in the bearings and running under normal conditions. The ends of the shafts are fitted with balancing rings. The machines are operated at full speed and balance weights are added to minimize any detected vibrations. Traditional balancing techniques employ instrumentation such as accelerometer mounted on the machine structures. These are effective for reducing vibrations in the rotating equipment near the points where the sensors are attached.

The fiber optic approach yields information about the bearings rather than the machine structures. It is somewhat less useful than use of structure-mounted sensors for balancing, but the rotational frequency vibration measurements made with fiber optic monitors can indicate clearances between bearings and housings while systems are operating. Such effects are generally undetectable at the outer surfaces of the rotating machines. The data can determine whether clearances are adequate to prevent the bearings from becoming locked in the housings due to dimensional changes that occur with thermal gradients. The clearance measurements also show whether the bearings have been jammed in the housings during installation.

Noise Level

Noise generated by geometrical imperfections in balls or rollers and raceways is often used to indicate overall bearing quality. The noise is generally measured with an Anderometer (Ref. 16) and expressed in anderons—RMS vibration amplitudes taken over bands of 50-300 Hz, 300-1800 Hz, and 1800-10,000 Hz. These bands isolate and quantify wearing surface roughness parameters such as eccentricity, waviness, and surface finish. A military specification establishes vibration limits for bearings based on Anderometer tests (Ref. 17), and serves as an absolute standard for controlling bearing quality.

Anderometers can check bearing quality before installation or after removal from machinery. Fiber optic monitors can extend the concept to in-service components.

Noise levels are lower in machines than on Anderometers because of constraints imposed by installation housings. Vibration levels of installed bearings, determined using fiber optic displacement probes, are highly correlated with Anderometer measurements. It has therefore been possible to define maximum vibration limits for bearings in-service on the basis of an absolute numerical anderon specification.

Bearing Defect Factor

Broad-band RMS vibration, measured as bearing noise, indicates irregularities uniformly distributed on wearing surfaces. Dents, scratches, inclusions, or spalls do not contribute significantly to RMS values. These isolated singular imperfections cause impacts that generate narrow peaks in the bearing displacement signal. A bearing free of isolated defects generates a displacement signal with a uniform amplitude distribution across the frequency range of interest.

An instrument sensitive to peak energy can be calibrated to indicate defect factor. The minimum defect factor is unity for a bearing without isolated flaws. Values above 1.5 indicate defects that can shorten bearing life.

Figure 2 shows defect factors for a highly loaded bearing that developed spalling. Initial readings of 1.1 indicated freedom from defects. By the end of the second week, values of about 1.65 indicated that a defect had formed. Using pattern recognition on the signal (Ref. 10), it was determined that the defect was on the inner ring. After ten weeks, the failure rate accelerated and the system was stopped. The bearing was disassembled, revealing three spalls on the inner ring, as predicted. This failure was detected by the fiber optic method nine weeks before increasing vibration levels would have been noted using other techniques (Ref. 18).

Bearing Speed Ratio

The ratio of the outer ball or roller pass frequency to the rotational shaft speed is proportional to the operating loads and clearances. Speed ratio limits based on normal variances of operating loads and clearances can be calculated and used as criteria for installation and operation. Measured deviations from predicted ratio can indicate abnormalities such as improper loading, faulty mounting, and unsatisfactory lubrication (Ref. 12)--any of which can lead to early failure.

Fiber optic probes provide signals for direct calculation of the bearing speed ratio. For moderately or heavily loaded bearings, the speed ratio can be directly measured and compared to tabulated values. For lightly loaded bearings, tunable bandpass filters can be incorporated in the systems to extract the ball or roller pass signals.

APPLICABILITY

Because fiber optic sensors are small, lightweight, passive, and insensitivity to electromagnetic interference, they are suitable for applications that have previously defied instrumentation. The ability to sense conditions within a rotating bearing is an example relevant to many industrial, commercial, and military systems. Fiber optic probes for microsurgery offer other illustration. Increasing use of these devices for monitoring and control can be expected.

REFERENCES

1. "Method of Evaluating Load Ratings for Ball Bearings", AFBMA Standards, Section 9, Oct. 1960, Anti-Friction Bearing Manufacturers Association, Inc.

2. Tauber, T.; "A New Chip Detector--Reliable, Versatile, and Inexpensive", Proceedings of the 23rd Meeting of Mechanical Failures Prevention Group, National Bureau of Standards Special Publication 494, pp. 123-132, Sept. 1977.

3. Senholzi, P.B.; "Tri-Service Oil Analysis Research and Development Program", Proceedings of the 22nd Meeting of the Mechanical Failures Prevention Group, National Bureau of Standards Special Publication 436, pp. 165-184, Dec. 1975.

4. Scott, D.; Wescott, V.C.; "Predictive Maintenance by Ferrography", Wear, Vol. 44, 1977, pp. 173-182.

5. Collacott, R.A.; "Vibration Analysis", Vibration Monitoring and Diagnosis, Halsted Press (New York, NY), 1979.

6. Dyer, D.; Stewart, R.M.; "Detection of Rolling Element Bearing Damage by Statistical Vibration Analysis", ASME Paper 77-DET-83, Design Engineering Technical Conference, Sept. 26-30, 1977.

7. Darlow, M.S.; Badgely, R.H.; "Application of High Frequency Resonance Techniques for Bearing Diagnostics in Helicopter Gearboxes", USAAMRDL-TR-74-77, US Army Air Mobility Research and Development Laboratory (Fort Eustis, VA), Sept. 1974.

8. Drago, R.J.; Board, D.B.; "High Frequency Vibration Monitoring Techniques for Gear/Bearing System Failure Detection", AGMA Paper 109.36, 1973 AGMA Meeting, Montreal, Canada, Oct. 1975.

9. Fischer, J.P.; "Marine Ball/Roller Bearing Monitoring by Shock Pulse Monitoring", Proceedings of the 32nd Meeting of the Mechanical Failures Prevention Group, National Bureau of Standards Special Publication 622, pp. 126-136, Oct. 1981.

10. Philips, G.J.; "Fiber Optic Machinery Performance Monitor", US Patenr 4, 196, 629, 8 Apr. 1980.

11. Philips, G.J.; "A New Technology for Bearing Performance Monitoring", Proceedings of the 22nd Meeting of the Mechanical Failures Prevention Group, National Bureau of Standards Special Publication 436, pp. 18-28, Dec. 1975.

12. Philips, G.J.; "Bearing Performance Investigations Through Speed Ratio Measurements", ASLE Transactions, Vol. 22, No. 4, Oct. 1979, pp. 307-314.

13. Philips, G.J.; "Rotating Machinery Rolling Element Bearing Performance Monitoring Using the Fiber Optic Method", Reliability, Stress Analysis, and Failure Prevention Methods in Mechanical Design, ASME Publication No. H00163, 1980, pp. 37-43.

14. Lagace, L.; Kissinger, C.; "Non-Contacting Displacement and Vibration Measurement Systems Employing Fiber Optics and Capacitive Transducers", Proceedings of the 23rd International Instrumentation Symposium, ISA, 1977, pp. 1-9.

15. Cook, R.O.; Hamm, C.W.; "Fiber Optic Lever Displacement Transducer", Applied Optics, Vol. 18, No. 19, Oct. 1979, pp. 3230-3241.

16. Chaney, L.; "The Anderometer", Mechanical Engineering, Aug. 1944, pp. 515-518.

17. "Bearings, Ball, Annular, for Quiet Operation", Mil Spec MIL-B-17931D (SHIPS), 15 Apr. 1975.

18. Nishio, K.; et al, "An Investigation of the Early Detection of Defects in Ball Bearings by Vibration Monitoring", 79-DET-45, ASME Design Engineering Technical Conference, Sept. 1979.

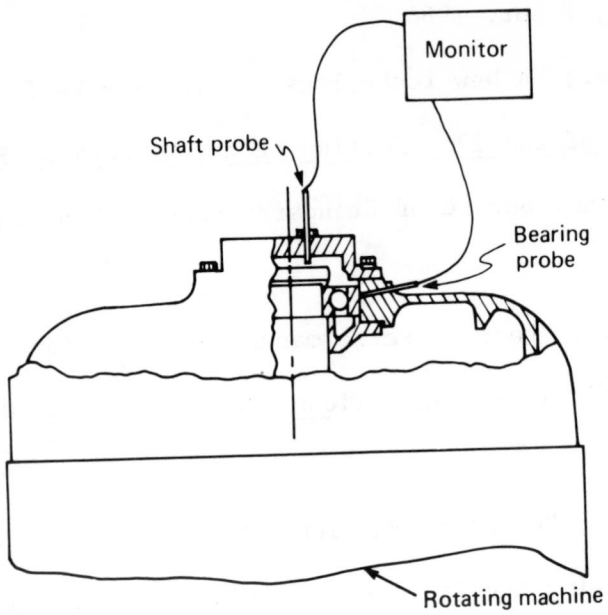

FIGURE 1. Typical fiber optic bearing monitor installation.

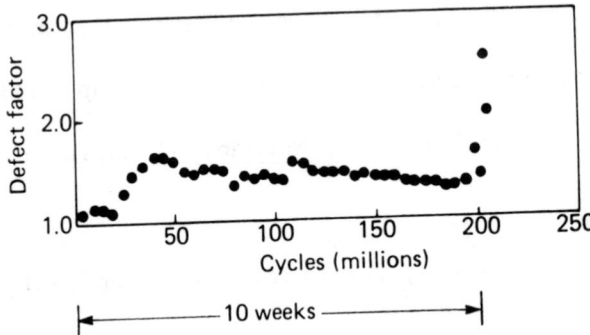

FIGURE 2. Bearing defect factor, as indicated by the fiber optic monitor.

CHAPTER 3

MAINTENANCE METHODS
AND PRACTICES

Reprinted with the special permission of *Dun's Business Month*, (formerly *Dun's Review*),
August 1979, Copyright 1979
Dun & Bradstreet Publications Corporation

The High Cost of Bad Maintenance

Last year, American industry spent more than $200 billion in maintenance costs for plant and equipment. According to maintenance specialists, at least $60 billion of that mammoth expenditure was wasted. And more than money went down the drain. Poorly maintained equipment produces poor-quality products, and equipment failures result in disrupted production schedules and delays in delivering orders. In short, bad maintenance management increases costs while decreasing product quality, productivity and customer service.

Why is industry wasting almost one out of every three dollars spent on maintenance? The experts place the blame squarely on top management. To many corporate executives, maintenance suggests little more than janitors sweeping floors and technicians in greasy overalls changing nuts and bolts. They have no knowledge of what maintenance entails or what it costs, and they are not interested in finding out. As a result, says consultant Michael Hora of A.T. Kearney, Inc., "Many top executives don't realize that maintenance costs can be controlled and that management systems can be instituted."

Actually, most maintenance managers and superintendents hold graduate degrees in engineering and are responsible for multimillion-dollar budgets. It is not uncommon for the maintenance manager in a large plant to spend, say, $100,000 just to fix one piece of equipment. But chances are that such expenditures will not come under any real scrutiny because no one else in the company knows what the money is being spent for or is able to figure out whether it is being spent wisely.

Without management control, costs pile up because of inefficient planning and scheduling. It is estimated that a repair job done on an emergency basis requires three times more manpower, time and money than a scheduled repair. Besides that, many maintenance costs are hidden--for example, the cost of shutting down a production line to repair equipment or the cost of lost production time and disrupted schedules.

Determining these costs is a tough job, the consultants say, because most plants do not keep adequate records of which pieces of equipment need repair or how much each repair costs, and very little analysis is done as to why a piece of equipment fails. "The economics of a good maintenance program shows up in increased utilization of equipment," says Vice President Thomas B. Foster of Emerson Consultants, Inc., "but unfortunately many companies keep very incomplete records of that utilization."

Inventory Mismanagement

The mismanagement of spare parts inventories is another costly item at many companies. One consultant tells about a steel plant where only half of $70 million worth of spare parts and equipment is inventoried. "Thirty-five million dollars worth of valuable equipment is just lying around in some storeroom or in some corner, and no one knows what is there or where it is," he says. According to Hora, that kind of situation is not at all uncommon. In fact, he says, most maintenance operations don't have any inventory system.

Also costly—and widespread—is lack of coordination between the maintenance operation and engineering and production. Maintenance crews often show up to overhaul a piece of equipment without first checking with the production manager about the best time for such repairs. More damaging, maintenance men say, they have a constant battle trying to convince production managers that it is sometimes necessary to overhaul a piece of equipment before it breaks down.

The major reason for these conflicts is that while the production manager is responsible for meeting his production quota and the maintenance manager for fixing equipment, chances are that no one is responsible for the cost of not maintaining the equipment. "As long as maintenance is viewed as a free service by the production people," Mike Hora says, "there is absolutely no incentive to hold down costs by doing preventive maintenance."

In such an environment, it is no wonder that the training and development of maintenance personnel is also being overlooked. Most maintenance department heads do not have management skills to start with, and they are seldom encouraged to develop them. Thus, the leadership vacuum goes right down to the department level, and the maintenance manager's job becomes a dead end. "The bright young engineer with aspirations of rising up the corporate ladder won't touch the job with a ten-foot pole," says one consultant.

Indeed, boding ill for the future, executive recruiters report that there is already a critical shortage of maintenance managers, particularly those with both engineering and management skills. "The situation can only get worse," says President Era Moore of Chicago search firm Cadillac Associates, "because you just can't find graduates willing to take the job."

To be sure, changing business conditions have been forcing companies to re-examine their maintenance operations in the past few years, and there is evidence of spotty improvement. The trend is particularly notable among chemical and mining and metals companies.

One reason for the growing attention is that the need for maintenance has increased as plants have become more automated and equipment more sophisticated. Besides that, with many companies running their plants at or near capacity and holding inventories to a bare minimum, unscheduled downtime resulting from a poorly managed maintenance operation has become increasingly intolerable. "With the cost of capital what it is today, a company cannot afford to have unused equipment lying around," says Frank Strang, a division general manager at heavy-equipment maker Rexnord Inc. "We used to have fallback equipment in case of failure, but that is a luxury we no longer have." As a result, Strang says, Rexnord has improved its entire maintenance operation in the past few years, which has led to less downtime and smoother production schedules.

Rising energy costs have been another incentive for companies to maintain their equipment at peak efficiency. At International Harvester Co., for example,

Ronald C. Reed, manager of facilities and energy, says that energy costs have risen 137% since 1973. "Maintenance is one area where we can make inroads in conservation," Reed explains, "because a well-running machine requires less energy than a machine or piece of equipment that needs repair." A well-running machine is also a safer machine, and stricter government safety regulations have been a further impetus to better maintenance.

New Emphasis

Not surprisingly, companies are coming to realize that improving management is the first step in making their maintenance operations more efficient. Minnesota Mining & Manufacturing Co., with its worldwide plant operations, is a good example. "We are changing the way we look at maintenance," says Joseph C. Juettner, director of maintenance and machine shops. "Rather than seeking maintenance managers or supervisors with technical backgrounds, we are now looking for businessmen who are capable of planning and managing people and projects, and we will supplement them with people who have a strong technical background."

Reynolds Metals Co. is one company that has long paid attention to maintenance. The company spent $250 million--almost 10% of sales--on maintenance of its operations last year, and every penny was well spent, according to President John E. Blomquist.

Reynolds stresses maintenance management. In order to attract and develop management talent, for example, it rotates managers among the production, engineering and maintenance departments. It also offers seminars in management techniques that are open to maintenance and production personnel as well as the traditional sales, marketing and finance people. Moreover, a stint in maintenance can lead to advancement. "Six of the eight managers of our major domestic operations have served in maintenance at one time or another," Blomquist says.

Another company, Alcan Aluminum Corp., recently held a seminar for its American and European maintenance, production and engineering managers to discuss proposed changes in the company's maintenance operation. "The emphasis has switched from fixing machines to keeping them running, and to that end we will be doing a lot more preventive maintenance as well as setting up planning systems," says Robert Ball, manager of one of the company's largest domestic plants.

Ball points to the company's Oswego, New York, plant as an example of the type of maintenance changes being planned for other operations. In Oswego, the maintenance manager, who previously reported to the engineering manager, now reports directly to the plant manager. Daily meetings between the maintenance and production managers have been instituted, and responsibility for ordering spare parts and equipment has been switched from purchasing to the maintenance department.

A number of companies have turned to computers to help predict when machines will need to be serviced. Inland Steel Co., for one, has computerized the service schedules for much of its equipment. Diagnostic equipment that measures mechanical vibrations and sound waves is also being used to determine when servicing is necessary.

Computers also help to keep track of inventories. At St. Joe Minerals Corp., for example, the spare parts and equipment in the giant storerooms at its various mining operations are hooked up to the purchasing department at headquarters by a central computer. "When the supply of an item gets down to a certain level, the purchasing department is alerted to reorder," says Lawrence Casteel, vice president of mining. "As a result, we wind up having to carry fewer supplies and are assured that we will not unknowingly run out of a critical part or piece of equipment."

Coordinating the maintenance operation with production and engineering is the most difficult challenge of all. Because maintenance has been ignored for so many years, production and engineering managers often do not realize that maintenance people can provide suggestions and ideas that could make the entire manufacturing process more efficient and less costly. As a result, they do not even ask for such input.

Utilizing Expertise

But some companies are making an effort to bring the various departments together. For example, if representatives from the maintenance department attend meetings in which production schedules are planned, a maintenance schedule can then be planned that will not interfere with production. Maintenance managers can also advise production managers about possible equipment problems that may arise under normal production operations; this gives the production planners a clearer picture of exactly how much they can expect to produce in a given period of time.

Maintenance managers can also be helpful with information when ordering a new piece of equipment--on its reliability, the estimated time and expense involved to maintain or repair it and whether replacement parts are readily available. Alcan is even bringing its maintenance managers into the early stages of equipment design. "Our maintenance people were recently involved in the initial stages of designing a crane," says Robert Ball, "and as a result of their suggestions, servicing that crane will take only twenty minutes, as opposed to two hours as it was originally designed."

In a decade or two, the consultants say, maintenance management may become as commonplace and as refined as production management is today. But for now, only a relatively few companies are giving it the attention it deserves. And the situation will not improve until top management understands its importance and makes the necessary commitment. As a maintenance expert at 3M quips: "I just wish someone would come up with a sexy term for 'maintenance.'"

--L.A.

Managing Maintenance for a Greater Contribution

By John D. Andrica
A. T. Kearney, Inc.

I – MAINTENANCE TREND

Over the past several years, maintenance cost, as a percentage of sales, has continued to rise. We're losing ground rapidly. In 1982, major companies reported maintenance costs that were from 2% to 16% of sales. Any cost that can be saved through better maintenance management techniques translate into direct profit dollars. Through better technology, training, and methods, maintenance can contribute to the overall health of a company, rather than just be a drain on its resources. However, maintenance is reported as an increasing cost in almost every industry that was surveyed during 1982. Industries reporting major increases in maintenance cost are:

Industries	Percent of Sales
Drugs	16%
Consumer Products	8
Automotive	4
Basic Materials	6
Chemicals	6
Engine Manufacturing	7
Foundries	8
Paper	10
Steel	15
Electronics	2

In addition to the increasing cost of maintenance, maintenance people are demanding higher wages and getting them. A sample of 200 plants indicates that the hourly maintenance craftsman wages are currently ranging from $8.00 per hour to over $17.00 per hour, with 36% of those plants surveyed paying over $13.00 per hour for a maintenance craftsman not including fringed benefits.

People involved in the management of the maintenance department are also demanding higher wages and getting them. In our 200-plant sample, we found that the typical first-line supervisor in maintenance was being paid between $24,000 a year and over $54,000 a year. Of the plants surveyed, 27% indicated that they paid over $35,000 per year for their first-line maintenance supervisors. The 200 plants that were surveyed also indicated that their maintenance managers were in salary ranges from $30,000 per year up to and including $64,000 per year with 50% of those surveyed indicating that they paid $40,000 or more per year to their maintenance managers.

The cost of maintenance for individual plants is reflected in the chargeback that is typically used within a manufacturing environment, where the maintenance department charges back to the user the cost of maintenance. That charge on the average was $33.00 per maintenance man-hour.

One of the major cost for the maintenance department is spare parts. While most maintenance departments indicated that the ratio between maintenance labor costs and maintenance spare parts was basically a one-to-one ratio, we found inventories within the 200 plants we surveyed ranging from $500,000 to over $3.5 million dollars. In America today, there is no doubt that maintenance represents a major investment.

THE BIG JOBS ARE GOING
GOING TO THE PROS

Even though the maintenance profession is catching up, and in many cases passing the production profession in terms of financial rewards, we are finding shortages of qualified management personnel to fill open slots in the maintenance managers rank. Recognizing that those maintenance manager jobs are paying $30,000 to $64,000 per year, companies are having an increasingly difficult time in filling those slots. The major concern that most companies are faced with is the need to have people in maintenance with management talent; mechanic ability is not enough to fill those jobs.

The lack of qualified applicants for the open managerial jobs in maintenance can be traced back to a time when people felt that maintenance was a dead-end area. Young graduates coming out of school looked to the production side of the house for career advancement. Most of our young managerial graduates or engineers with managerial majors would not touch a job with maintenance. Today production jobs, which are on a decline, have numerable applicants available for each slot, whereas in maintenance, where the bulk of the vacancies are, we're finding fewer and fewer qualified applicants to fill those spots. As a result, there is a definite leadership vacuum in the maintenance business.

II – THE RACE TO THE AUTOMATIC FACTORY

The armaments for the next industrial revolution are at hand. During the past few years machine toolmakers, many in the U.S. (even more in Japan) have begun to supply flexible manufacturing systems that herald something very close to the workerless factory. The repercussions of the new technology go well beyond the predictable improvements it brings to productivity.

Flexible manufacturing systems complete a process of factory automation that began back in the 1950s. First came numerically controlled machine tools that performed their operations automatically according to coded instructions on paper or tape. Then came computerized design and computerized manufacturing or CAD/CAM, which replaced the drafting board with the CRT screen and the numerically controlled tape with the computer.

The new systems integrate all these elements. They consist of computer control machine centers that produce complicated parts at high speeds and at great reliability, robots that handle the parts, and remotely guided carts that deliver material. The components are linked by electronic controls that dictate what will happen at each manufacturing sequence, even automatically replacing worn out or broken drill bits and other implements.

To make these systems function within a given factory environment will require two distinct technologies:

1. The manufacturing engineer who not only understands electronics and computers but is able to design products that are manufacturable on the flexible manufacturing systems.

2. The maintenance technician who is capable of repairing this new technology.

The maintenance profession will undoubtedly become the most productive profession within our new automatic factories; obviously uptime on this highly sophisticated equipment will be the final goal in making the new automation pay off.

THE BOTTOM LINE

Flexible manufacturing systems' greatest payoff lies in the capacity to manufacture goods cheaply in small volumes. The grandfather of today's approach toward manufacturing automation was the production line or what later became to be known as the automatic transfer line, which got its name from the transfer of the product being worked by a conveyer from one metalworking machine to another. The payoff of the transfer line was in high-volume, long-production runs. But such mass production is shrinking in importance compared with batch production from several thousand to one.

Seventy-five percent of all mechanic-machined parts today are produced in batches of 50 or fewer. Many assembled products are also manufacture in small batches. Even the automobile (the father of the mass-production line) is starting to give way to manufacturing lines that are flexible, which will enable the automakers to produce more low-volume models for small market segments.

SO WHAT IS THE COST

When compared against some of the machinery that the flexible manufacturing system replaces, they would appear, on the surface, to be extremely expensive. A full-scale system encompassing computer controls, five machine centers, and accompanying transfer robots can cost upward to $25 million dollars. Even a rudimentary system built around a single computer control turning center might cost about $325,000, while a conventional numerically controlled turning tool would cost only about $175,000.

But the direct comparison is a poor guide to the economies flexibility automation offers, even taking into the account the phenomenal productivity gains and production rates that come with virtually unmanned around-the-clock operation. Because a flexible manufacturing system can be instantly reprogrammed to make new parts of products, a single system can replace several different conventional machine lines, yielding high savings in capital investments and plant size.

The work force supporting our automatic factory will change in its composition. On the production side of the house, the people will be predominantly in the engineering staff and their daily duties or job functions will be more in terms of product design and product specifications, along with detailing automated approaches toward manufacturing. We will find few people out on the shop floor actually involved with hands-on manufacturing.

112

On the maintenance side of the house, we will be finding a more professional approach toward maintenance. The maintenance department will also be aided with a computer. The computer will track maintenance problems, maintenance cost, machining history, detailed preventive maintenance procedures and programs. The maintenance man will also be highly technical in his approach to maintaining this new technology.

How Is the Revolution Going

Within the United States, the new revolution dealing with flexible manufacturing systems or the automation of our manufacturing environment is going slow. In fact, 34% of the tools used in the United States for manufacturing are 20 or more years old. Less than 4% of our tools are numerically controlled, even though numerically-controlled machining tools have been on the market since the 1950s. In terms of Flexible Manufacturing Systems, the United States can only point to about 30 such systems currently in operation.

On the other hand, Japan is doing quite well in its quest for automating their factories; Japan only has about 18% of its manufacturing tools being older than 20 years. Toyota alone has more than 30 flexible manufacturing systems currently in use. However, awareness is spreading among U.S. manufacturers. The time is running short to reorganize production processes and to begin investment programs in new technology. Survey show growing interest in new machining tool purchases. Machining tool builders, still shocked by recession, are cautious but the new high technology companies are optimistic about their future markets. They expect sales of accessories alone for the robot, computer controls, and material handling systems to reach over $30 billion worldwide by 1990 compared with last year's total of only $4 billion.

Get Ready

As this technology becomes a way of life in today's industry, the people who today comprise the maintenance business must begin to get ready for this automation. You have to prepare to meet the challenge posed by this new technology. Obsolete managerial techniques and attitudes will not suffice in the automatic future. You can begin now to anticipate what your company will need in terms of maintenance to meet production demand on advanced technology. You can begin now to sharpen your managerial skills, to tighten your organization, and to train yourself and your staff to meet the technical requirements and demands that will be placed on you in the next few years. There is no doubt that maintenance will be the job of the future!

III - MAINTENANCE AND THE ACCOUNTING LEDGER

In the past, maintenance has been considered an unnecessary evil, overhead, something you just have to do if you're going to be in business. The main contributor to this perceived thought is that the maintenance department has always been considered on the wrong side of the accounting ledger. The department is usually considered an expense, not a source of income. But given that management's goals are always to keep expenses down and income up, maintenance is increasingly being recognized as a potential contributor to corporate profits by controlling costs.

To manage maintenance effectively and to assume a position as part of the management team within your company, you need to be able to talk in real dollars when you discuss money that you have saved the company. Maintenance is a business and you cannot inflate your contribution with hypothetical savings and still retain your credibility.

To determine your actual contribution to the profit of the company, you need to know:

1. Material cost - including inventory value, inventory turns, and annual costs.

2. Labor cost - both by craft and by area and by type of service provided.

3. Overhead cost - including supervision, administration, rent, fringes and supply items.

4. Downtime Cost - both to loss margin on sold out units and the variable cost on open time units.

When you know these costs, you will not only be able to discuss them and your operations intelligently, but you can also discover ways to trim fat and make your business more profitable.

Real savings can come from fewer people on the payroll, more production out the door with the same number of employees, or reduced overtime. These are qualifiable means to reduce cost and thereby generate actual dollar savings. The best way to attract these profit contributions is to work with and through the accounting department which reports in real dollars both by periods and by year.

Or better yet, measure your progress in terms of Assets Productivity. What is Assets Productivity? It can be identified as:

Time	Forward Thinking
Speed	Safety
Quality	Inventory Levels
Tool Life	Service Levels

and generally measured in terms of return on assets. The formula for calculating return on assets is:

$$\frac{\text{Net Income}}{\text{Total Assets}} = \text{RETURN ON ASSETS}$$

As the maintenance department's impact on equipment reliability and uptimes become more constant, there should be a marked reduction in product inventory levels, and an increase in overall asset productivity.

IV - MAKING PROFIT DOLLARS IN MAINTENANCE

Before you can improve the performance of your department, you have to measure it. This is the first step in making maintenance a profit center. To measure performance, you have to collect detailed information about the:

1. Percentage of downtime compared to other companies (this information can be gathered from trade journals and societies, and industrial averages).

2. Percentage of maintenance job performance based on standards.

3. Percentage of emergency work.

4. Percentage of stock outs.

This information can help you plan. Planning and estimating your needs can save your department as much as 20% of its present costs. As a manager, you need to establish standards of performance, a formalized work order system, and positions that allow your people to use their time, money, and techniques to the department's best advantage. You need to know both the maintenance department's history in each of these areas and an accurate prediction of future needs. In short, this is where you begin to answer the question, "What should I do to generate more profit?"

MEASURING TECHNIQUES

Although the most complete way to measure labor performance is through engineered standards, these are also the most expensive. Simplified maintenance standards will probably give you sufficient means to evaluate work performed, at a much lower cost. Simplified work standards are essentially a basic average of what you can expect from your labor force. They are formally defined as:

Elemental breakdowns of maintenance or construction jobs into direct work and auxiliary work. The application of standard times to these (job elements) and the subsequent totaling of these times comprise a job standard.

The advantages of having performance measurements are valuable to you in tightening the productivity of maintenance. Because they are consistent and tested, standards allow you to measure performance accurately and to identify conditions, such as poor equipment and facilities, that may detract from the performance of your workers. In addition, they provide a means to improve operating methods, served as sound basis for scheduling, and provide backlog control.

This is not to say that standards do not have any drawbacks. For one thing, they incur relatively high administrative costs. They must also be updated and maintained regularly, which requires a clear definition of procedures, organization, and systems.

Let us examine how performance standards can influence direct work, work such as cutting pipe, digging holes, painting a wall, or removing a pump. To measure the quality of the laborers' performance, you compare the actual time used to finish the job against a standard time. If, for example, your standards indicate that a specific job should require two men for four hours (or eight man-hours) and it is actually completed by three men, working five hours (or 15 man-hours), then the performance of the crew is only 53%. Clearly, such standards help you identify weak peformance. Your goal should be from 60% to 80% consistent efficiency.

SAVINGS OPPORTUNITIES

In addition to developed standards, there are several ways that you can improve savings and reduce costs. One of these is formal maintenance scheduling, by week and by day. Maintenance, production, and planning should work together to schedule personnel. This method can save as much as 5% of your costs.

Maintenance can save yet another 5% through storeroom control, which would include an inventory control system, locked gates, and parts kitting and delivery. A maintenance information system, which produces accurate and timely reports on manpower distribution, foreman performance, planner performance, job status, and equipment histories by exception, can save your company another 5% in expenditures for maintenance.

Although there is no way to measure the savings in real dollars generated by preventive maintenance, this is still an area which requires your attention. Equipment identification, a P.M. service list, a history on the equipment, and reports evaluating the results of your programs can all contribute to the overall program of cost reduction. As part of your preventive maintenance efforts, you should be able to identify both the frequency and the man-hours required to keep a specific machine functioning smoothly. It might, for example, require 100 man-hours per year to inspect the thickness of vessels or eight man-hours per month to inspect a crane. You have to know this information for all the jobs within your department so that you can accurately plan for future manpower, time, and financial requirements.

There are benefits in real dollars that result from using some or all of the suggestions discussed here. A large chemical company in the Midwest, for example, implemented a new work order system, a computerized information system, and work standards. While reducing downtime from 8% to 2%, the maintenance department saved the company $1.5 million in hard dollars. In another case, a paper company in the South used similar programs to increase labor performance by 14%, while decreasing mechanical downtime to 2%.

V – WELCOME TO THE PLANT MANAGER'S STAFF

Although you may already be on the plant manager's staff, do you really participate? The programs to save money outlined above can make you an active member of the management of your firm. In fact, to be accepted as part of the management team, you have to contribute to profit, savings, and capacity. And, not only do you have to be active in controlling maintenance costs, you have to be able to discuss the improvements you have made in real dollars to gain the recognition you deserve.

Because you provide capacity, you should be involved in production scheduling and in determining the amount of capital spent on upgrading production capabilities. If, for example, your company decides to install robots, you'll need to know how much maintenance they will need and how much that will cost. This information is necessary to make intelligent decisions about whether or not to use robots, how much, and in what areas.

One way to ensure that you get the kind of information you need to contribute to such management decisions is the computer. Only the computer can quickly and accurately provide data on cost, labor utilization, equipment activity, downtime, performance, exceptions, actual performance versus goals, and adherence to schedule. And, the computer can give you these data regularly.

If you want to be more than an engineer, if you want to be an active part of management, you have to be accountable - you have got to have the facts, to get attention, and to make money. The future will require professional mangers. Entry level jobs in maintenance should be filled with potential managers for the future, because growth tomorrow depends on what we do today.

Management skills for you and your staff will be required and can be gained through training. Although an MBA may not be necessary, some formal training will position you for management responsibilities. Last year, U.S. industry spent from $10 to $40 billion in maintenance - this represents a real management opportunity. Maintenance is a big business now - and its future belongs to the professional manager.

Reprinted from *Production Engineering,* Copyright Penton/IPC Inc., April 1978

Selling Management On Downtime

THE DOWNTIME DILEMMA

By JOHN H. MCRAINEY / *Executive Editor*

Can management policies *cause* downtime? The answer is probably, "yes." But the question is unfair; "cause" is too strong a word. But it is not unfair to ask whether management can be a part of the downtime problem. The answer is definitely, "yes!".

Consider a two-act tableau that is played out again and again in all types of manufacturing companies: In a desk-thumping scene, an irate manager is vowing that equipment will be made to operate and a production goal will be met—or else! In a second scene, the same manager is reviewing cost sheets, or a purchase requisition, or a budget; as he scrawls "disapproved" on the paper he is

heard to mutter something about one-dimensional engineers who don't realize that a company is in business to make a profit and not to goldplate a production operation.

The two scenes highlight a serious shortcoming in the prevalent practice of management: The manager attempts to pursue simultaneously goals that conflict with each other. In the context of

this article, the manager attempts to maximize equipment uptime and, also, to minimize the costs of both equipment and equipment maintenance. It can't be done—and most likely neither goal will be met. And worse! In trying to achieve conflicting goals, a manager ends up creating discord and confusion among various functions within his plant where teamwork and focused effort are really needed.

This tendency of managers to stress conflicting goals and its unhappy effects were evident in the answers to a recent survey of PRODUCTION ENGINEERING's readers. Downtime was a noteworthy problem in most plants, affecting in the final analysis things of top concern to managers—timely shipments, cash flow, customer goodwill, etc. Yet, when asked, "How could management better support your efforts at reducing downtime?", almost all responders indicated something that their management could or should be doing but wasn't. The summation of the various indictments is that operating groups simply are not getting the type of informed management support they feel is needed to really solve downtime problems.

Some of the lack of support probably reflects not ignorance, but a management willingness to gamble and hope (despite Murphy's law) that what can go wrong, won't. But it is apparent from the survey that there are many higher managers who do not understand or appreciate either the complexities or subtleties in obtaining and maintaining a healthy production operation.

Problem areas

Reflecting the differences in plants across a variety of industries, survey answers about management shortcomings fell into three broad categories:

Purchasing policies—Apparently too many companies still focus on the initial cost of things they buy rather than the total cost of owning them. Engineers are only too aware that the inexpensive, "competitive" purchase that created a hero in the purchasing department yesterday can in the future haunt an operating department with trouble, again and again.

The survey responses noted that the false economy in purchasing by price alone applies not only to equipment but also to associated control components and maintenance spare parts as well.

Some responses called for improvements in qualifying suppliers. Here the implication is that purchasing is aware that something other than price should be considered in a technical buy—but that purchasing policies don't do a good job in identifying that "something." The basic problem is a lack of coordination between purchasing and the shop floor. In short, the survey revealed clearly the not-too-unreasonable attitude of those on the production floor that, if they are to be responsible for keeping equipment producing, then they darn well want their knowhow and experience to be reflected in selecting exactly what is used. The point is so obvious it is surprising that there are still companies in which engineers find it necessary to emphasize such a need.

Operating policies—The survey results suggest that managers and engineers in some plants have conflicting views of production equipment. Evidently the man-

---

WHAT'S NEEDED
Here's how management can help reduce downtime

- ■ Realize that complexity of modern controls has changed the plant environment and definition of what's normal and abnormal.
- ■ Realize that every thing new and modern is not an improvement so far as operations are concerned. Conversely, some older, still-working items should be replaced with vastly improved things now available.
- ■ Maintain better records of causes of failures, cost of repairs, cost of lost production.
- ■ Provide more technical support for in-house R & D to update purchase specs and to qualify suppliers; zero-in on fewer sizes, more select control items. Buy items by other than price.
- ■ Draw operating experience into purchase of equipment and controls—and provide for better incoming inspection.
- ■ Establish critical classification of spares and stock accordingly.
- ■ Recognize value of control features and stand-alone instrumentation that help sense, pinpoint, and correct problems.
- ■ Realize that equipment scheduling and use are critical factors in the downtime problem. Allow and plan for periodic overhaul and maintenance.
- ■ Allow time to actually correct a problem rather than demanding a quick fix.
- ■ Understand some minor problems should be corrected immediately even if a line must be shut down, instead of waiting for an absolute failure.
- ■ Don't attempt to rush new equipment and processes into production use before development and debugging are completed.
- ■ Provide adequate training for operators and maintenance people. Train operators to note and report slight operating irregularities. Push to have operators and mechanics work together in making repairs.
- ■ Pay for outside specialists to check critical equipment periodically. Support periodic recalibration of critical, sensitive sensors and control elements.
- ■ Establish "downtime reduction program" based on a team from production, design, maintenance, and other affected functions.

ager is primarily concerned only with the quantity of shippable products turned out now, in the present. The engineer concedes the importance of output but he cannot ignore technical realities—equipment has a finite life, it can be abused; and almost inevitably, little problems that are ignored grow up to become big problems.

Engineers see that production goals will best be served if more managers have more of an engineer's view of the operating production line—by not overloading the system, by providing proper operator training, by making equipment available for maintenance and overhaul, and by not demanding maximum output from equipment before it is properly broken in and debugged.

In the same vein, engineers want managers to realize that some trouble and downtime are inevitable—and that some can be foreseen. The engineering solution to such problems is to be prepared to handle them as expeditiously as possible. This requires training of both operators and maintenance people and a stock of spare parts—particularly for critical machines and activities that really control a plant's output. It also includes a willingness to spend more for equipment with built-in diagnostic aids, and the realization that those aids are not "frills," but mean shorter downtime in the future.

A new perspective—Some survey responses reflect another kind of difference between managers and engineers immediately involved with the production floor. The picture that emerges is one of managers who simply haven't taken the time to assess the impact of change in the plant environment.

Engineers would have such managers understand that new, more complex equipment and controls cannot be introduced into a plant without a rethinking of the way many functions within the plant relate to the production line.

Observers of the production scene have long preached that au-tomation and more complex controls would stress the maintenance function. Much higher skills would be required to keep machines up and running. That point has been very evident in plants where tasks once allotted to setup men or mechanics are now performed by staff engineers and where any equipment trouble requires an engineer's attention.

But some engineers apparently now sense the need for something more than specialists and higher-skilled individuals to replace the jack-of-all-trades types in mainte-

nance. In addition to higher skills, a higher organizational approach is required. In a sense, the downtime problem has moved out of the "call-a-mechanic" stage and into the "convene-a-committee" stage.

The essential point is not really that a committee be formed but rather a realization that in a systems sense, the general downtime problem has moved from the simple fix-it level to a higher level worthy of broad, detailed managerial attention. The choices are many—a new equipment acquisition can be the best solution to a problem; frequent and detailed backup service by equipment suppliers can pay dividends; changes in product design can help keep production machines running; a change in suppliers of purchased parts can improve output; a major overhaul and refit can be more economical than continual erratic equipment performance.

It isn't easy to change policies

Applying such solutions, and more, clearly depends upon how a plant's management views the downtime problem. The broad, systems viewpoint is fundamental to achieving real changes since it would automatically lead to a reconsideration of purchasing and operating policies. If a plant's management lacks this systems viewpoint, operating personnel and engineers are faced with one of the more difficult, thankless problems in the industry: helping someone higher up in the organization to see the light.

Selling ideas upwards can be real battle—and, as in battle, victory turns upon things like allies, weapons, strategy, tactics, timing, . . . and luck. The downtime problem should make natural allies out of the production engineering, operating, and service functions. But get as many different functions involved as possible— design, production control, quality control, accounting, purchasing, and marketing have a legitimate

CONTINUED

interest in a well-functioning production operation.

The best weapon is cost data—get the real cost of downtime, not the typical accounting department figure in which the costs of specific events are hidden in summary accounts. The real cost of a specific downtime event includes lost production, scrap, time of all people involved in the fix, value of replacement parts, etc. This total cost also becomes an item in a log of the downtime history of the machine and of the cause of failure or by some other classification.

In time, this type of cost data

should capture the eye-opening horror stories that would impress management. But more can be done. Any plant with a variety of equipment probably contains examples of poor equipment design that contribute to maintenance problems—hidden lubrication points, maintenance access doors that can't be reached without some machine disassembly, filter bowls that can't be observed, etc. These types of problems should be photographed and worked up into a report as a pointed reminder that production floor experience cannot be ignored in the equipment purchase. If a recurring cost figure can be associated with such design goofs, so much the better.

Note here that the horror stories and various memorandum reports are used initially just to plant and nurture the thought that a worthy

problem exists—not to support a suggested line of action. In short, concentrate on selling the notion of the problem before attempting to sell any solution.

Once management has been convinced that a real problem exists, there is a built-in commitment to solve that problem. But if management is presented a single package that attempts to sell both the existence of a problem and a solution, attention will be drawn to the solution details. And if those details are lacking in some respect the proposed solution can be rejected—and the idea of the problem as well.

Part of the strategy in helping management see the light, then, is to decide when the ground is sufficiently prepared for a formal proposal that a policy change be made. When the time is ripe, take your best shot in a crusade against downtime.

With recognition of the problem and a commitment to seek a solution, the organization takes over responsibility for solving the problem. Your task now is to help tailor that solution—it must be viable and satisfy the realities of the organization and its various power centers. And if you have done your homework, yours can be the dominant voice in discussing the detail pros and cons of suggested changes that can make management a part of the *solution* to the downtime problem—and not a part of the *problem*.

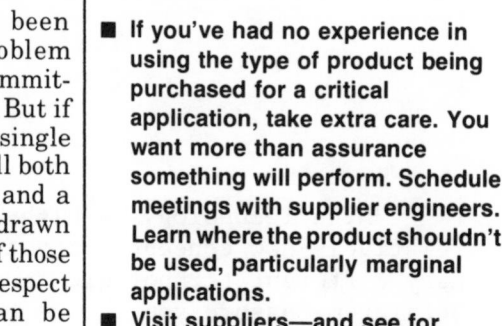

Helpful hints for purchasing quality performance

- If you've had no experience in using the type of product being purchased for a critical application, take extra care. You want more than assurance something will perform. Schedule meetings with supplier engineers. Learn where the product shouldn't be used, particularly marginal applications.
- Visit suppliers—and see for yourself just what effort they go through to create and deliver a quality product.
- Investigate the supplier's policy in handling hurry-up orders for spare parts and engineering assistance. Are there nearby sources of spares?
- Pick up a typical service manual from the supplier and have your own operating and maintenance people examine it. Is it clearly written? Can it be followed? Does it suggest some problem area or source of trouble that hasn't been considered? Does it describe how to interpret those little signs of probable forthcoming trouble with the product?
- Some suppliers suggest certain kits of spares be purchased with the product. What's included in the kit? Why those items?
- Various suppliers provide maintenance contracts for their products. The terms are of obvious interest. More importantly, find out what the supplier has experienced in servicing such contracts—meaningful statistical data should be available.

Presented at the CASA/SME Autofact 5 Conference, November 1983

Multicraft Maintenance Management in a High Technology Environment

By Michael J. Wood
John Deere Component Works

I. INTRODUCTION

"High Technology" – High Tech – is the premier buzz word of today, a relatively new concept not yet fully developed which actually encompasses many different technologies and means different things to different people. But let me assure you this – to the "old school" maintenance craftsmen and supervisors that still work in your factories and mine, "High Tech" still means the same old thing – "fix it right, faster and at a lower cost".

This paper will discuss the challenges of managing a multicraft maintenance organization in a high-tech environment. Specifically, I'll share with you the approach used at the John Deere Component Works in organizing and operating our maintenance groups as we enter the high-tech era. I'll be covering in detail such items as –
* Organizational Structure and Objectives
* Management Tools/Communication Tools
* Engineering Support
* Using High Tech to Maintain High Tech

The John Deere Component Works is a large manufacturing complex situated on an 800 acre site near the center of Waterloo. There are approximately seven million square feet under roof and nearly 5,000 machine tools in service with an average age of over 20 years. The machine tool technology ranges from simple milling, drilling and turning machines to complex computer-controlled FMS and robotic applications used to manufacture major drivetrain, hydraulic and special components for agricultural and industrial vehicles.

II. MAINTENANCE GOALS

There are two primary goals in the maintenance organization at John Deere:
* Minimize Maintenance Cost
* Improve Machine Availability/Uptime

Over the last ten years, we at John Deere, have seen a dramatic increase in the cost of maintenance, (as a percent of sales) not unlike most other industries. And since the cost of maintenance adds no value to the product, any dollar saved through improved maintenance management is an additional dollar of profit to the company.

The important point to note here is that these goals – minimizing cost and improving machine availability – are the same goals we had five years ago and will probably have five years hence. The goals have not changed. What has changed – and is continuing to change – is our approach to achieving these goals.

We've come to see that maintenance must come out of the steam tunnels and press pits and adopt a much more professional strategy or approach to accomplishing it's objectives. It's our belief that through better management, methods, training and technology, we can accomplish just that.

III. MAINTENANCE ORGANIZATION

Given this philosophy, we chose to begin our programs at the beginning - with our organizational structure. Our maintenance organization at the Component Works is structured as shown in Figure 1. The structure itself is not unlike any traditional management hierarchy. What we have done however is adopted an Area Maintenance strategy to provide machine repair services to the manufacturing departments. We have a Machine Overhaul/Support Services group which provides a pool of craftsmen to be used by the Area departments in times of peak workload in addition to providing machine overhaul, component rebuild and other specialized services.

We also have a consolidated Maintenance Machine Shop which provides critical back-up support to the Areas and which, in itself, utilizes High Technologies in performing it's functions.

It's two of these areas of our operation - the Area Maintenance groups and the Maintenance Machine Shop - that I'd like to expand and focus upon throughout the balance of the discussion.

IV. AREA MAINTENANCE

The Component Works factory is serviced by five area maintenance departments whose boundaries were determined primarily by geographic area (square feet) and type of manufacturing (assembly, heat-treat, cast iron machining, etc).

Each Maintenance Area is staffed with a Supervisor, Maintenance Planner/Scheduler, Material Coordinator and a multi-craft crew consisting primarily of electricians, mechanics and toolmakers. This crew is responsible for all Preventive/Predictive Maintenance, Scheduled Maintenance Repairs and Reactive Breakdown Repairs within their area. The advantages to us of the Area Maintenance concept over a centralized maintenance group have been:

A. Reduced Maintenance Costs -

Travel time between jobs has been greatly reduced as can readily be seen in a factory as large as the Component Works.

The crew's efficiency has increased since the smaller groups allow the craftsman to become much more familiar with a smaller number of machine tools.

These two items alone have significantly reduced machine repair costs and helped in improving manufacturing productivity.

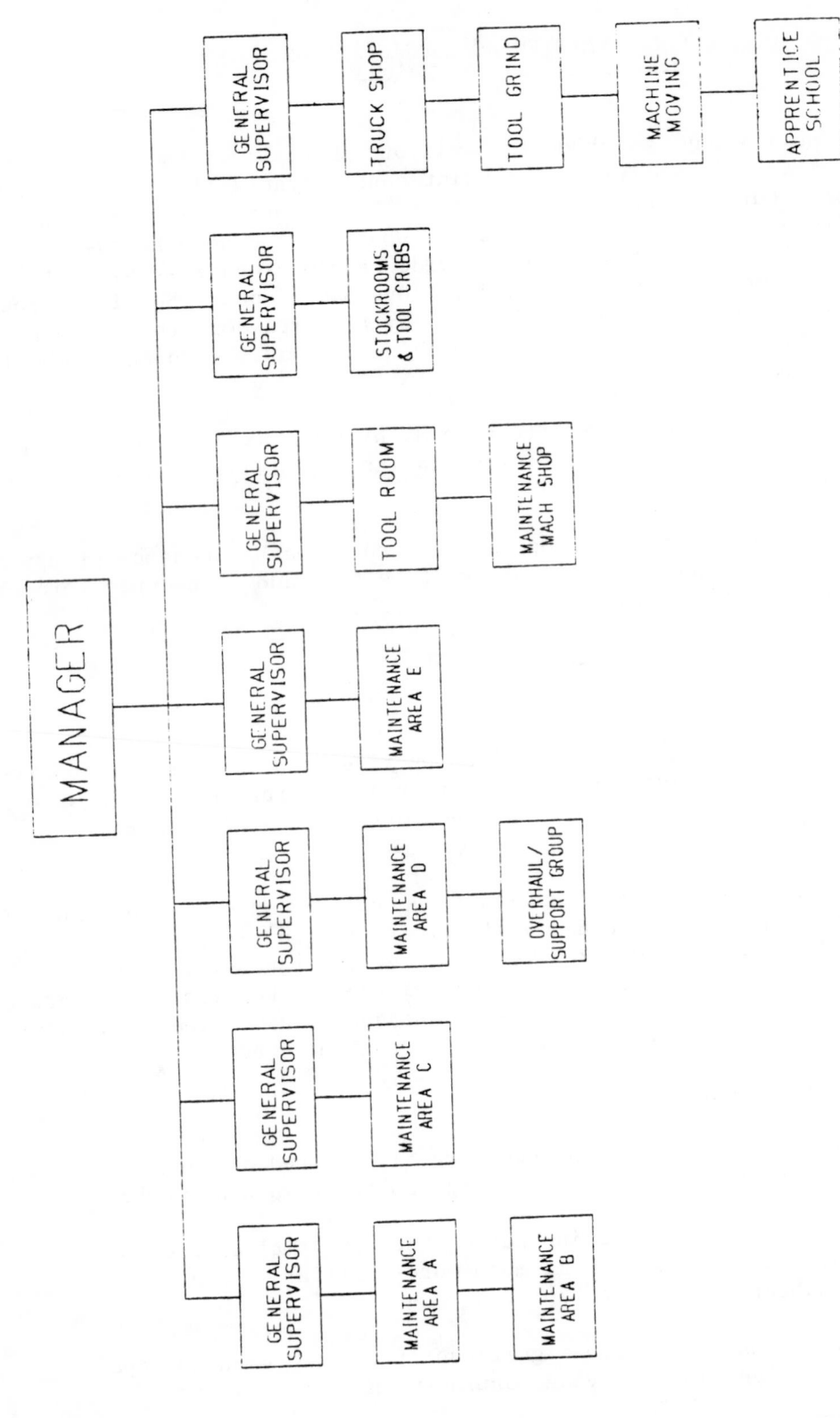

MECHANICAL SERVICES ORGANIZATION
FIGURE I

B. Be Able To React Rapidly To Emergencies -

Being located centrally within the geographic areas they serve, our maintenance areas are now able to react very rapidly to emergency situations. Within the last 12 months alone, the average response time to a machine breakdown has been cut in half.

C. Communication Between Production and Maintenance -

The Area Maintenance concept has put the maintenance supervisor out onto the shop floor with his craftsmen - precisely where he belongs. This has led to increased visibility and awareness on the part of both the manufacturing and maintenance supervisor and has helped to improve working relationships immensely.

D. Able To Eliminate Chronic Problem Areas -

The central geographic location, smaller coverage area, increased visibility and the assistance of his Maintenance Engineering team - which will be discussed in detail later - has allowed the maintenance supervisor and his crew to focus on chronic problem areas. We've seen great reductions in problem areas and complete elimination in many cases.

Within each Maintenance Area works a Maintenance Planner/Scheduler and a Material Coordinator in direct support of the Supervisor and Craftsmen. These individuals are the backbone to success of any Maintenance Area. These individuals work in close coordination and are responsible for:

* Preparing work schedules and holding planning sessions with Manufacturing and Engineering personnel.

* Inspecting jobs and estimating labor and material requirements.

* Keeping everyone informed of job progress.

* Checking availability and obtaining materials and parts required for repair.

* Coordinating fabrication and shop repairs.

A relatively new feature within the maintenance organization has been the installation of On-Line W.O.R.K.S. (Work Order Record Keeping System). This is a Maintenance Information System that allows direct, on-line communications between engineering, the manufacturing department and the maintenance department. Eliminated are the handwritten work orders which invariably get lost in the company mail and the phone calls which never get answered. Also eliminated is speculation on how long this or that machine has been down, how frequently it breaks down, how many manhours were spent, the cost of materials and so on.

What is gained by this system are such things as:

*   On-Line Machine History including type, location, age, specifications, maintenance cost (labor and materials), etc.

*   Work Order Request and Assignment.

*   Workforce Planning and Backlog Reports.

*   Machine Availability.

*   Job Prioritizing/Scheduling Features.

*   Job Status/Pending Job Reports.

*   P.M. Schedules.

*   Ability to Maintain and Update P.M.'s., machine history files, etc.

*   Labor and Material Usage and Cost Data.

This system has provided a Return On Investment in excess of 100% and is fast becoming the premier communications and management tool within the maintenance organization.

Complete, accurate and timely documentation of course is a requirement for adequate maintenance performance. Each Maintenance Area is equipped with a Microfiche Reader/Printer and a file containing microfiche documentation on all critical machine tools located within the area. Included in the microfiche files are service manuals, parts manuals, electrical diagrams and other prints and information. This eliminated ragged, tattered and lost documents and saves valuable time spent looking for information. A well-defined procedure also assures that microfiche are updated or replaced as necessary. It's anticipated that microfiche of all machine tool documentation will soon be available within all Maintenance Areas.

Mentioned earlier was a Maintenance Engineering team supporting the Maintenance Supervisor. Within each of the five Maintenance areas is physically located a team of Mechanical and Electrical Engineers to work in direct support of the Maintenance Supervisor and his crew.

This team of engineers has four basic ongoing tasks:

A. P.M.'s -

A significant amount of their effort (as high as 60%) is devoted to the development and enhancement of Preventive/Predictive Maintenance Schedules within the Maintenance Area. Close contact is maintained with Manufacturing, the Maintenance Supervisor and the craftsmen during this process to assure that the P.M.'s are credible, timely and valuable.

B. Technical Support -

Secondly, the team exists within the area to provide technical
support to critical maintenance problems. In this capacity, the
engineer may be providing direction to the supervisor or craftsman,
assisting in obtaining or correcting documentation or engineering,
or providing liasion contact with service representatives.

C. Engineering Solutions -

Thirdly, the Maintenance Engineer assists in the development and
implementation of engineering solutions to chronic or longstanding
maintenance problems.

D. Training -

Lastly, the Maintenance Engineer is responsible for determining
technical training requirements within the Maintenance Area and
for developing and conducting necessary training sessions for all
maintenance personnel.

The Area Maintenance concept has proven to be quite successful at the Component Works, however, it would fail for certain were it not for the critical
support of the Maintenance Machine Shop and Support Services groups.

V. MAINTENANCE MACHINE SHOP

Prior to the formation of the Area Maintenance groups, we operated with
three separate, distinct machine shops. Each shop had it's own equipment,
people and areas of responsibility. In addition to wasting valuable floor
space, equipment and personnel, it also was very confusing to the customers
requiring service.

Having recognized the problems that existed under this type of operation,
we proceded with a plan to combine the three machine shops into one machining
complex. The obvious benefits of consolidation materialized and today we operate with less total square footage and fewer employees.

The formation of a large machining complex, however, created a significant
job tracking problem. The increased volume of work orders and material flowing
throughout the department during a three-shift day resulted in scheduling difficulties and work occasionally being lost or misplaced.

The installation of a DEC PDP-1134 mini-computer based scheduling system
aids our planners and supervisors in prioritizing, estimating, scheduling and
tracking work flow through the department.

All work coming into the Machine Shop is received by the Planner who checks the job requirements and enters all necessary information into the mini-computer where a maintenance work order is opened and a copy is generated on a printer. A copy is given to the person requesting the work and a second copy travels with the job through the shop. When a machinist is assigned a job he is required to "log on" the computer which aids in tracking the job through the process.

This system allows the Planner to recall any job from the computer for inspection or updating. Ready access to stored information has virtually eliminated past problems of losing jobs and has allowed the generation of backlog reports, pending work, "Hot" lists, etc.

In addition to consolidation and the mini-computer, several new pieces of equipment have been installed which have rapidly advanced our Machine Shop into the "High-Tech" era.

A new machining concept recently installed is the M.D.I. Programmable Lathe. This is not an N.C. machine using a pre-programmed tape. Rather, it has a microprocessor control which is programmed by the machinist. Once the data is put into the machine and the part has been run, the program can be stored on cassette for future use.

The part shown in Figure 2 is an example of the type of job which is ideal to run on the M.D.I. Lathe. Turning this part on a conventional lathe would take approximately 4.5 hours. The same part on the M.D.I. Lathe takes about 12 minutes, excluding the time for programming.

We also have installed one wire EDM machine in our shop and are finalizing plans for the installation of two more. This computer-controlled machine cuts with an electrified wire through virtually any kind of conductive material regardless of hardness or density. This machine has opened up a whole new realm of machining capabilities.

The on-board computer allows small jobs to be manually input to machine while tapes are generated for larger jobs with more complex geometry or for repetitive jobs of long duration. A temperature controlled room compliments this machine's ability to hold tolerances and repeatability of $\pm$ .0002 in material up to 4 inches thick.

Many previous machining operations such as milling, slotting and jig-boring can now be done with the Wire E.D.M. after the piece has been heat-treated. By elimination of heat-treat distortion, maintaining tolerances and mating fits is virtually assured.

Presently the type of work normally done on this machine includes tracer templates, die punches and inserts, exotic repair parts and prototype work.

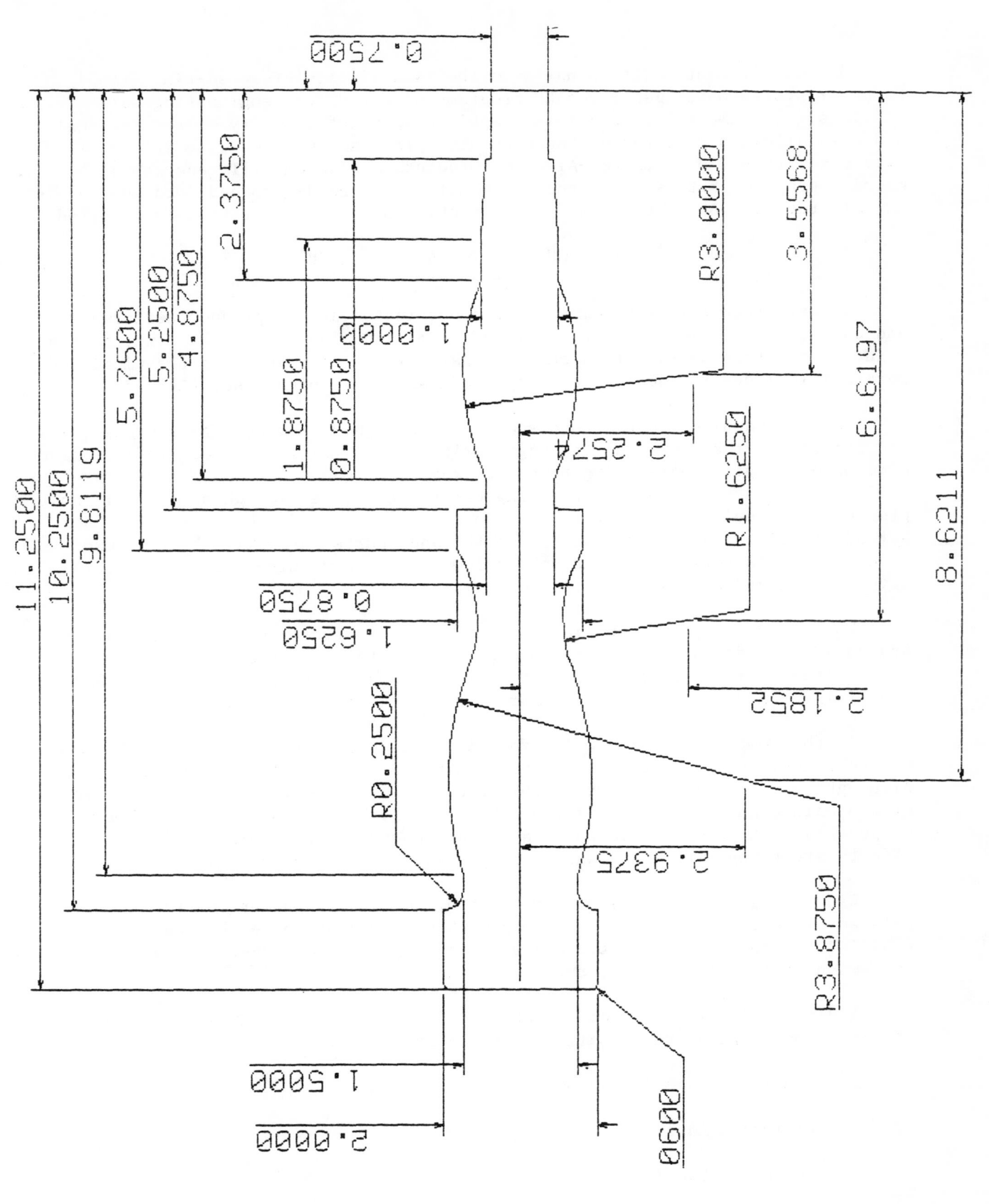

Our most recent addition has been the installation of a Cad-Linc Graphic Computer system which generates NC programs directly from engineering data or drawings. The GNC software allows graphic display of geometry on the terminal and also allows simulation of the machining process. Coupled with it's tooling data base, this CAM (Computer-Aided Manufacturing) system will generate a machine program that is right the first time with pre-determined feed and speeds. And all this can be accomplished in a fraction of the time previously required.

VI. "USING HIGH TECH TO MAINTAIN HIGH TECH"

As our product line becomes more sophisticated and as product performance and quality become more directly related to maintaining market share we, in the maintenance organization, find our responsibilities becoming more technical and more critical and the quality of our workmanship more highly scrutinized.

We find machine tools on our shop floor designed to manufacture parts to specifications that were only an engineer's dream a few years ago. We also find some older machine tools being pushed beyond their capabilities because of stricter tolerancing. We're producing parts with geometries whose names I can't pronounce and we're producing to tolerances a fraction of the thickness of a strand of hair. Terms like "half-a-tenth" and "parts per billion" are commonplace in our factory today when only a few years ago machine alignment within several thousandths was adequate.

What this all points out is that the crescent wrench and ball-peen hammer are no longer adequate in maintenance of machine tools of today's complexity. We've already discussed the need for improved management tools, better methods and training and now we have need for improved technology in maintenance.

At the Component Works, we're focusing the introduction of new maintenance technologies, for the present, in our Machine Overhaul/Support Services group. Although most of the resources of this group are still directed to the support of the maintenance areas, traditional machine overhauls and the rebuilding of components such as hydraulic pumps and small powered hand tools significant efforts are being directed toward improved maintenance technology.

Within the last few years, for example, we have begun our own repair of printed circuit boards from production machines. We now have the equipment, training and personnel to test, evaluate and repair almost every circuit board on a machine tool in our factory.

Additionally, we've recently introduced into machine maintenance the use of three tools which truly qualify as High-Technology:

A. Laser Technology

B. Vibration Analysis

C. Infrared Scanning

The use of laser technology in machine maintenance has multiplied rapidly since it's introduction within the last couple years. All new machine installations are checked for flatness and alignment as well as linear motion. Checks for repeatability, backlash, endplay or ballscrew wear are commonplace. The laser also allows checking where virtually impossible using traditional methods as in checking pitch and roll on transfer lines up to 100 feet long or in verifying rack alignment 80 feet high and 400 feet long in our high-rise storage areas. The use of the laser has clearly made critical checking easier, faster, more complete and more accurate than traditional methods.

Vibration analysis is being used extensively in our P.M. schedules in identifying stress points caused by poor alignment and improper balance. This is a significant aid in preventing premature component failure particularly in bearings, belts, shafts and motors.

An infrared scanning device is also getting increasing use in P.M. schedules in the detection of hot spots on a machine in operation which leads to areas of potential failure.

VII. CONCLUSION

We've progressed a long way in a short period of time toward meeting the new challenges of the high-tech environment. We've definitely climbed out of the steam tunnels and press pits. The management, methods, training and technology - all requirements for a successful maintenance organization - have been identified and put into place. Where we are exactly along the road I'm not sure but I do believe were definitely heading in the right direction.

What do we see in the future? Definitely we'll see increased use of computers in scheduling, tracking and monitoring activities. We'll see further developments and improvements in diagnostic, checking and measuring tools. We're experimenting now with the use of statistical control techniques in developing predictive maintenance. Beyond that, who knows? Consider the possibilities. . .

Reprinted from *Commline*, November-December 1980

REDUCING

by Howard Cooper

This article will cover in detail how the use of maintenance graphs, histories, and NC maintenance logs over a six year period have helped companies predict causes of high NC maintenance downtime. These predictive indicators were then used to develop NC maintenance programs for the elimination of such causes. The sequential evolution of these programs will be covered in detail including the areas of: training NC technicians, establishing an organized spare parts inventory program, establishing a flexible electronic repair lab, establishing a practical preventive maintenance program (both electrical and mechanical), establishing an electronic controls standardization program, and finally a program to eliminate adverse environmental conditions for NC and industrial controllers. These programs can result in maximum machine tool up time and maximum dollar return. Individual reports on each of these programs will be published in upcoming 1981 issues of COMMLINE magazine.

Data Collection and Reports

First, let's look at the specific kinds of reports and maintenance logs which were used by us in collecting statistics and predictive indicators. Then, we will look at how the information was used in collecting these statistics and predictive indications. Finally, we will look at how the information was used to develop useful programs for reducing maintenance downtime and increasing NC utilization.

An individual NC machine tool log was kept at each machine in a special steel pocket made for that log. At the top of each page of the log are the headings: today's date, name of maintenance personnel, time problems, and solutions. These headings remind the maintenance personnel to give the date of each downtime or service call as well as the time of day, the maintenance person's name, the problems which summoned him to the machine, and what the corrective actions were, which put the machine into service again.

The Production Department Supervisor should keep a chronological machine status report as shown in Figure 1. This report is for a 24 hour period. Horizontally the machine tools in a given department are listed and vertically the times-of-day in which the status of each machine tool changed are listed. Notice that the last machine tool on the right hand side changed status more times in a 24 hour period than did the others, therefore, the last summary line is marked after the last status change at 11:20 p.m. of this machine tool. Also notice that at 11:30 p.m. which is the ending and beginning of each day the status of each machine tool is forwarded from its last status change to the 11:30 p.m. listing. A numerical listing of the status definitions can be found in Figure 2. In Figure 1 we can see that the second machine tool was in the productive mode at 11:30 p.m., status code 45, and continued productive until 7:05 a.m. at which time status code 40 indicates the machine went down for electrical maintenance and it remained down until 10:08 a.m. At that time it was put back into the productive mode and continued being productive from 10:08 a.m. until 11:30 p.m. which is considered the end of the day.

Manufacturing

Engineering can then take these records and summarize them into both a weekly and monthly downtime report, similar to that shown in Figure 3. This report is broken up into the three main departmental areas, which may occupy the machine's time. The production department codes are at the top, maintenance downtime codes are in the center, and the tooling department including NC programming codes are in the bottom section. By breaking the report into these three sections the utilization breakdown can be quickly seen. 702 productive hours or 84% of the machines' time was accounted for by production. This does not mean however the machine tool was up and running 84% of the time. Notice that on the top line which is indicated by code 45 (productive time), the total at the right side

of that line indicates that for all 7 machine tools in that department the actual total productive time was 407 hours or 48% of the weekly total. The production department must also take the responsibility for non-productive NC utilization in those categories listed in the top section, i.e.; as the first machine tool was down for 24 hours or 20% of the week, due to set up status 20. The third machine tool was down for 24 hours because there was no material, this accounted for 20% of this machine tool's total week. The fourth machine was down for 64 hours or 53% of the week due to the operator being away from his machine and the first machine was down for 56 hours or 47% of the week because the operator was not in the plant.

In the maintenance section of Figure 3 are some of the most valuable statistics; here we see that the fifth machine tool was down for 20 hours or 17% of the week waiting for parts (electrical), the second machine was down for 20 hours waiting for maintenance (mechanical). Notice that these two blocks of time which were attributed to the maintenance department were not hours spent repairing the machine tool but were direct inefficiencies of the department, either waiting for parts or waiting for personnel. Therefore, the ratio of maintenance efficiency can be directly picked up from this chart. Compare the wait times versus the number of hours spent in actual electrical or mechanical maintenance, as indicated by status code 40 and 41. It can be seen from the right hand column of the maintenance section that for this group of machine tools, maintenance spent far more time getting their ducks in a row. Also, notice that the maintenance downtime as a total for this department was 138 hours or 16%. These numbers are then turned into line graphs for each machine tool, for each department, and plant wide graphs showing the three graphing lines, one for production, one for maintenance, and one for tooling.

Other helpful statistics were taken later as a result of organizing a pre-

NC DOWNTIME

DEPT. 10 DAILY REPORT OCT. 4, 1980

MACHINE NUMBERS

MACHINE NO.	25166	99830	12694	01372	14386	13113	20148	20150	20151
Time	1130P	1130P	1130P	1130P	1130P	1130P	1130P	1130P	1130P
Status	45	45	20	31	47	45	27	40	41
Time	700P	705A	208A	333A	705A	930A	130A	127A	635A
Status	40	40	45	45	41	23	40	45	45
Time	936A	1008A	700A	410P	405D	1205P	705A	436P	710A
Status	26	45	40	27	28	39	27	39	27
Time	115P		900A	635P		329P	1030A		715A
Status	45		36	40		45	40		40
Time	431P		1125A	640P			1040A		800A
Status	27		44	45			45		45
Time			1205P						910A
Status			25						37
Time			130P						1000A
Status			32						45
Time			230P						1230P
Status			45						28
Time			800P						208P
Status			61						41
Time									1120P
Status									45
Time	1130P	1130P	1130P	1130P	1130P	1130P	1130P	1130P	1130P
Status	27	45	45	45	28	45	45	39	45

To see the daily report previous to this one type D again. This can be done up to seven times so a report can be received for each day in the past week. If monthly or weekly averages are needed see the next two sections, 2.d. and 2.e..

Figure 1

Possible Codes

Machine Down for:

20	*P	Setup
21	*M	Waiting for electrical parts
23	*T	Tape tryout and editing
24	P	Special study code
25	P	Operator away from machine
26	P	No operator in today
27	M	Waiting for Maintenance—Electrical
28	M	Waiting for Maintenance—Mechanical
29	P	Inspection (waiting for layout)
30	P	No material
31	P	Reclaim
32	P	Warm-up
33	P	Miscellaneous delays
34		
35	T	Tooling tryout
36	T	Perishable tool troubles
37	T	Fixture troubles
38	M	Waiting for Maintenance parts (electrical)
39	T	Waiting for part tape
40	M	Electrical Maintenance
41	M	Mechanical Maintenance
42	M	Preventive Maintenance (Electrical)
43	M	Preventive Maintenance (Mechanical)
44	P	Part Handling
45	P	Machine in production
46	P	Methods
47	M	Waiting for parts (mechanical)
48	P	Available but not run

*P —Production
M —Maintenance
T —Tooling and Programming

Calls:

61	Call Shop Supervisor
62	Call Programmer
63	Call Collant Pumpout Man
64	Call Machine Oiler
65	Call Electrical Maintenance (machine running)
67	Call Electrical Maintenance (machine down)
68	Call Mechanical Maintenance (machine running)
69	Call Mechanical Maintenance (machine down)
70	Call Job Kitter
71	Call Relief Operator
72	Call Shop Area Man
73	Call Shop Scheduler
74	Call Chip Tub Man
75	Call Method (fixturing)
76	Call Tooling

Figure 2

ventive maintenance program. These will be shown later as the preventive maintenance program is discussed.

Real Time Reporting

It should be pointed out here that of the above records kept, the graphs are used most by upper management, the tables and charts are most useful to front line supervision, and the individual machine tool logs or histories are critical to technical personnel whose responsibility is to determine the cause of repeated downtime and eliminate it. It should also be mentioned here that a great expense can be wasted generating such reports if adequate management and technical time is not spent evaluating the result of these reports. In large shops where there are 15 to 20 or more NC machine tools, it may become necessary to implement an automated system which reduces the number of human hours required in keeping these statistics and which would totally eliminate human hours needed in compiling these statistics and generating the reports. There have been several systems developed in the past few years, such as the Plant Maintenance Control System by IBM's Clinton H. Johnson, Jr.; the Digital Equipment Corporation Plant Management system which offers real time data to Managers; the Boeing Corporation Help System; the Ford System; the Sundstrand Maintenance Reporting System which works in conjunction with their DNC Systems; and the CICLOPS System, developed by this writer. A good manual reporting system for plants with fewer machine tools was reported by the magazine Machine and Tool Blue Book in an article written by Walter J. Reid, Managing Editor, entitled "NC Reporting Plots Course To Manufacturing Efficiency." Let's now see how these reports actually help to reduce machine tool downtime.

Predict and Prevent

When this writer first went to work for the John Deere Dubuque Tractor Works in 1974, the position filled was created as a result of extreme electrical and electronic downtime on the NC

DEPARTMENT STATUS REPORT—WEEKLY

DEPT. 10 WEEK ENDING OCT. 9, 1980

PRODUCTION Code	Description	25166 HRS	%	99830 HRS	%	12694 HRS	%	01372 HRS	%	14386 HRS	%	13113 HRS	%	20148 HRS	%	ALL HRS	MACH %
45	Productive T	40	33	88	73	32	27	80	67	42	35	90	75	35	30	407	48
20	Set Up	24	20											25	21	49	6
29	Inspection																
30	No Material					24	20									24	3
31	Reclaim																
32	Warm up																
33	Misc Delays																
34	Cleanup																
24	Spcl Stdy Cd																
25	Oper Away							64	53	32	27					96	11
26	No Oper In	56	47	32	27	40	33			etc						202	24
44	Part Handling																
46	Methods																
48	Avail Not Run																
	TOTAL	120	100	120	100	96	80			etc						702	84
MAINTENANCE																	
21	WFP Elect									20	17					20	2.4
47	WFP Mech											30	25			30	3.6
27	WFM Elect													18	15	18	2.1
28	WFM Mech			20	17											20	2.4
40	Elect M									26	22			20	17	46	5.5
41	Mech Maint			4	3											4	.5
42	PM Elect																
43	PM Mech																
	TOTAL			24	20					46	22	30	25	38	32	138	16
TOOLING																	
39	WFPT																
23	Tape Tryout																
35	Tool Tryout																
36	Perish T Trbl																
37	Fixture Trbl																
	TOTAL	0	0	0	0	0	0	0	0	0	0	0	0	0	0	0	0
ALL AREA TOTALS		120	100	120	100	120	100	120	100	120	100	120	100	120	100	840	100

Description Abbreviations:
WFP—waiting for parts
WFM—waiting for Maintenance
Elect M—Electrical Maintenance (actual active time spent on machine)
Mech Maint—Mechanical Maint. (actual active time spent on machine)
WFPT—waiting for part tape

Figure 3

machines. No downtime percentages are specifically referred to here because any given percentage of downtime, such as 1%, 40%, or 5% downtime has a very diverse meaning and implication from one plant to another. This is due to the different methods of tracking machine tool utilization. However, the downtime figures that this writer started working with were so high that typical comments were, "the mechanical downtime we can live with, but the electrical downtime is putting us out of business." Or, comments such as, "with the electrical downtime being more than five times greater than mechanical downtime, something has got to be done."

Training Program

With this predicament in mind and having looked over the reports and maintenance logs, it became most apparent that the biggest factor contributing to electrical downtime was training. Even though these machines work automatically, they do not repair themselves automatically. Therefore an extensive NC Maintenance training program was implemented whereby the electrical maintenance engineer would attend the specialized maintenance training classes taught by the machine tool or control vendor. The engineer would then train one or two NC technicians or electricians at a time after returning from school.

The recent boom of electronic technology into industrial controls has been so wide spread that in just a few short years, it has totally replaced conventional electrical control methods. Therefore our electrical repairmen must be completely replaced or retrained.

In 1948 both industrial controls and electronic controls were heavily based on relay logic. In that year the first computer, ENIAC (electronic integrator and calculator), was built with relay logic. For the next twenty-five years, industrial controls underwent very little change, while computer and electronic technology were continually advancing, each generation of technology was built on the last; from relays, to tubes, to transistors, to integrated circuits, to LSI, to mini-computers and now microprocessors and one chip micro-computers. Each generation was technologically more complicated and more powerful than the last, but because of the smaller size, less power consumption and mass production, the component cost continually spiraled downward. Each succeeding generation lasted a fewer number of years than the last, because each increased the momentum of advancement. When the integrated circuit came of age it was so small, powerful, easy to use, and inexpensive that industrial controls could no longer stay with relay logic and compete. The turning point came in 1972, when industrial control designers widely accepted the electronic technology. Therefore, during the past eight years integrated electronic technology has spread into all industrial controllers, even simple inexpensive machine tools. During that time industrial controls have progressed along with electronic technology from integrated circuit to large scale integration, to mini-computer, to micro-processor, not to mention OP AMPs, D/A (Digital to Analog), and A/D (Analog to Digital) modules. If an electrician has not worked on a new machine tool in the past five years his knowledge and skills are completely outdated by at least three generations. The gap between relay control and solid state circuit control, and the gap between IC control to mini-computer and microprocessor control are each as wide as the gap between horse and buggy and the motor cars of today. Reliable statistics indicate that possibly half of today's electricians will never cross the gap. Also, it is a definite fact that no electrician will cross these gaps *unless he receives extensive retraining* and experience nearly equivalent to a new apprenticeship.

The solution to this problem was to establish an in-house training program utilizing an electronic engineer instructor whose duty it was also to work with the electrician on trouble-shooting numerical control problems. By using this method the engineer was familiar with his students and their skill levels. Also, he naturally made his training directly applicable to troubleshooting needs as his alternate duties were to help the electrician troubleshoot those systems.

As the training program proceeded from basic training to detailed and in-depth training on specific numerical control systems, it was noticed that the downtime graphs and charts improved greatly on groups of machines which were controlled by the specific NC controller which had just been covered in training.

Spare Parts Inventory Program

As the electricians got a handle on trouble shooting techniques, the downtime record still showed large amounts of time waiting for parts. Therefore an extensive program was initiated to review needed electrical spare parts and to purchase and properly inventory such items so that they would be readily available to the repair electricians.

Test and Repair Facility

As the spare parts were put into use it was noticed that the downtime decreased slightly, but the maintenance logs indicated that many times the spare circuit board needed was out of the shop being repaired from a previous use and this repair turn-around-time was excessive in the area of one to three months. Even in cases where these spare modules were expedited at great expense, it still took the better part of a week to get the needed parts in plant. Therefore the next program was put into effect, that of a flexible, capable circuit board test and repair lab. This test and repair lab very quickly replaced out-of-plant repair and OEM exchange on circuit boards, tape readers, power supplies, and 98% of all electronic modules. The in-house repair turn-around-time was on the average twelve hours versus one to three months and if a board was expedited usually two hour repair time was possible. This test and repair lab not only reduced our need for multiple spare parts of a particular type, it also greatly increased the technical understanding and abilities of the NC maintenance force. They were now able to repair the NC systems in minimum time, having spare parts and then able to repair the defective module off line in the test and repair lab to the component level. This became very profitable from two aspects; Number 1, some $300/module was being saved over the OEM repair costs, and number 2, the machine tool downtime again decreased. This test and repair lab consisted of some $40,000 in test equipment. $20,000 being for general purpose electronic troubleshooting equipment and another $20,000 on very specialized digital troubleshooting equipment for digital circuit boards.

Predictive Analysis Helps

By this point in time the electrical downtime problem had been solved to the satisfaction of many in that electrical maintenance downtime had now been reduced to a level of being approximately equal with the mechanical downtime the plant was suffering. This bought enough time for those involved to step back and look at the information which had been collected and to analyze what was causing the majority of remaining downtime. This predictive analysis brought about the establishment of three new programs, the Preventive Maintenance Program, the Control Standardization Program, and the Environmental Control Program.

Electronic Controls Standardization Program

There were many new controls being purchased and it was noticed that often times there was no thought as to which electronic controller might best fit into a particular department due to similar controls already in that department causing reduced operator training time and reduced maintenance training time. Therefore a program was initiated which resulted in specifications of three standard numerical controllers and in addition, a list of standard SCR axis drives and spindle drives, and programmable controllers. The list of standard recommended controls remained very useful for a number of years. It avoided one of a kind odd controls and also avoided some misjudgment on new controls as the technology evolved from CNC through distributed microprocessor control to high speed microcomputer CNC controls.

Preventive Maintenance

Simultaneous to the standardization program came the implementation of a practical preventative maintenance program. It is quite common that few companies can dedicate enough time to preventive maintenance to accomplish all of the recommended PM steps which are specified by the OEM vendor. However, there are many practical preventive maintenance steps which save great amounts of downtime as well as repair expense: such as cleaning tape readers, keeping air filters cleaned and replaced, keeping critical lubrication levels high, replacing DC motor brushes for eight dollars rather than waiting until the brush wears out and the coil spring wraps around the motor commutator, thus requiring the replacement of a $2,500 DC motor, etc. This PM program concentrated on those parts of an NC system which are known to deteriorate with time and usage. Step-by-step check list procedures were written for each machine tool and the preventive checks were taken on a weekly basis with more detailed and extensive checks being taken on a quarterly to a semi-annual basis. The main strength of this PM program, to achieve continued success, was built into an auditing system whereby the operator and production supervisor monitored the checks made by maintenance and maintenance audited or monitored the checks made by the operator and the production supervisor. Also, the electrical maintenance engineer monitored the entire program and made predictive maintenance analysis from the information received. This information was fed back to the engineers by means of the PM procedural check sheets.

Power Line Transient Suppressor and Environmental Control Program

It had been noted for sometime that large amounts of our total NC downtime was attributed to a small number of the machine tools which seem to have consistent downtime for one reason or other. When the PM program was instigated we then turned to the PM feedback and the Servo Data histories to find what might be the common cause of this repetitive downtime. Close analysis of the available information indicated three things which were later substantiated by closer monitoring of the controls. These three things were heat build up in the control cabinet on a few machine tools, vibration and intermittent contact problems on a few other machine tools, and in most cases the problem was related to power line transient activity disrupting the electronic components and information flow within the control. Up to this time these problem machine tools had been identified and many of them were giving such excessive downtime that a control retrofit was being seriously considered. Some of these controls were of the older hard wired style, but some of them were very new CNC's. Regardless, the repetitive downtime and circuit board failures were making a complete control retrofit seem most desirable. However, upon discovering the three above environmental problems and developing an effective program to eliminate these problems, the machine tools and the controls were brought back to full reliability without a control retrofit. All of the subject CNC controls responded excellently to transient and temperature control. The hard wire controls responded equally well and in some cases special attention had to be placed on controlling vibration and using special contact solution on pin connectors and board edge connectors, as well as installing the power transient suppressors. Only two of these machine tools were ever retrofitted. On these two machines the electrical reliability had been established, however because of the limited control features these older controls had and because the machine tools had mechanically deteriorated to such a stage that they could no longer hold tolerances, the decision was made to carry through with the retrofit. The point to be learned here is that many NC and CNC controls are subject to

About the Author

Howard Cooper was born in Vernal, Utah and raised in Southern Utah. He is married and has four children. He has a BS degree in Electronic Technology from Southern Utah State, graduating Magna Cum Laude and has done graduate work at BYU. He has worked for the past nine years in Manufacturing Industries specializing in Industrial Electronics. Howard worked five years for John Deere Company as Numerical Control Maintenance Engineer and has worked the past year for Envirotech Eimco PMD as Electronic Engineer. In these positions he has been in charge of training NC technicians and developing effective maintenance programs on some 100 NC & CNC machine tools, and other electronic, industrial controllers. He has also been honorary member of the National Advisory Board for Electronics magazine.

high maintenance downtime and often times become candidates for retrofit, when really the only problem is one of power line transient reoccurring damage or one of heat and vibration environmental problems. When this was learned and an appropriate program setup for monitoring and correcting these problems on all NC machine tools, the electrical maintenance downtime dropped to an all time low of being something on the order of ½ the mechanical downtime, even though mechanical downtime had decreased due to fewer electrically caused failures. Electronic circuit board failures on those machines which were transient suppressed became almost non-existent, and we no longer had the memory problems and executive loss problems often associated with CNC machine tools.

A program was then established to protect all NC machine tools and electronic industrial controllers against power line transients and to properly protect the few controls which had heat or vibration environmental problems. To be effective, transient suppressors must be 6 nano second response time, 16,000 joule, soft clamping in combination with RF filtering capability to clean up the sign wave pattern. You should also get consulting to insure proper configuration and physical location for the suppressors.

Conclusion-Recommended Sequence of Programs

When this writer went to work for Envirotech (the Eimco Process Machinery Division) one year ago, I reflected back on the above experiences with this main question in mind, "In what sequence should these programs be implemented to provide Eimco PMD with the greatest results in the least amount of time?" After a period of getting close to the new machine tools and getting a feel for causes of downtime, it was determined that most plants have similar environmental conditions and the causes for downtime are generally the same from one plant to another. It was then realized that the *sequence* in which the above programs had evolved was *exactly backwards*.

That is to say, if any industrial plant would first install transient suppression to properly protect all industrial controllers and then survey to find over heating conditions and vibration problems and correct them, this would immediately eliminate most electronic and electrical failures. Then if a practical PM program was established the NC downtime would become so low that it would be hard to justify extensive programs for spare parts stocking and in-house circuit board test and repair. Although this writer feels training is a must for NC and electronic technicians, with the environmental problems cured and preventive maintenance active, the trouble shooting requirements would be reduced to a level that training would not have to be on such a rushed basis and would not be such a critical factor. With this in mind a survey was taken and capital expense justification was immediately written for transient suppression equipment and a few air conditioners. While this justification was in the paperwork mill for authorization, training classes were taught on basic solid state and digital electronics, also in-depth NC trouble shooting classes were conducted on a few of the specific NC controls owned by Eimco. During this time much of the engineers' time was spent trouble shooting problems with NC technicians both on regular time and overtime.

In June of 1980 we lost 50% of our NC technician force and in July, during the plant shutdown, the suppression equipment and two air conditioners were installed. Since that time the NC technicians have worked no overtime and very little straight time on NC machine tools. Even though their force has been cut in half the engineer's schedule has been eased to a major extent which has enabled him to work on training programs, establishing preventive maintenance procedures, and other engineering projects for manufacturing engineering. Since July we have had only two circuit board failures, one of which was an infant mortality failure on a new CPU board and the other board failure was directly linked to the explosion of a plasma cutting torch which caused an explosion power surge of 500 volts at some

2,000 amps.

In conclusion we should list the main points which were learned, in proper priority.

1. Predictive maintenance can be very helpful in determining and eliminating sources of NC downtime.

2. *Do not* spend a lot of time and money setting up predictive maintenance surveys until your equipment has been surveyed and properly protected from environmental impact. Namely power heat, vibration, and transients. This is inexpensive and the most cost effective thing that can be done.

When this writer goes out to survey a plant it takes him only four hours to survey 8 to 10 machines and 8 hours for 20 to 30 machines. A report is then written specifying the exact suppressor types and air conditioner types to be used along with the specified electrical connection points and physical location. The cost to properly protect an electronic controller ranges from $500 to $2,000 depending on suppressor configuration and air conditioning needs. This is a one time cost and can easily be justified with the elimination of the first 24 hours of downtime.

3. Predictive maintenance and machine tool log histories are invaluable in setting up a practical preventive maintenance program and for stocking spare parts, which will realistically be needed.

4. An effective training program should be established. Training proves very effective in reducing diagnostic time on NC machine tools.

5. If you have large numbers of electronically controlled machine tools it may be very advantageous to establish an in-house test and repair facility to reduce spare part inventories and module repair turn around time and expense.

6. If your company is purchasing numerical control machine tools frequently, it is most advantageous to establish a standardization program which will ensure like controllers within a given department or area and also reduce spare part inventory necessary.

Reprinted from *Production Engineering*, Copyright Penton/IPC Inc., October 1980

Weight the Averages to Lighten Downtime

By DILEEP G. DHAVALE
*Associate Professor of Production
and Operations Management
University of Wisconsin—Parkside
Kenosha, Wis.*
and
GEORGE L. OTTERSON, JR.
*Works Manager
John Deere Engine Works
Waterloo, Iowa*

In 1976, Deere & Company began making large diesel engines at its 1 million ft² John Deere Engine Works in Waterloo, Iowa. Products are for equipment such as farm tractors, harvesting machines, construction equipment, and irrigation systems. Over 500 machine tools produce the engines. They include transfer lines, dial, boring, and milling machines built especially for the Engine Works, and many older machines transferred from the Waterloo Tractor Works. The composite is a capital-intensive factory that must operate efficiently if the Engine Works is to make optimum use of its assets. The mandate: Find an effective way to allocate maintenance resources to maximize equipment uptime.

Objectives were laid out for the major areas of Productivity and of Maintenance Planning, much like those that any similar capital-intensive plant could target.

These are the Engine Works' Productivity objectives:
■ Operate equipment at optimum capacity and efficiency, especially the more expensive equipment.
■ Reduce equipment idle time, operator overtime, and scrap.
■ Improve Maintenance Department effectiveness.

The Maintenance Planning objectives:
■ Improve efficiency of labor usage.
■ Reduce unexpected breakdowns.
■ Reduce downtime from breakdowns.
■ Improve maintenance scheduling.
■ Improve efficiency and administration of preventive maintenance programs.
■ Maintain process capability to insure part quality.
■ Maintain equipment performance.

Find, rank, and weight

The method the Engine Works selected to meet those objectives is easy to use, yet easily modified to fit requirements and conditions of different plants. Because most of the machines at the Engine Works were new, historical data requirements were low.

Here's how the method works: First, identify factors affecting the Maintenance Planning and Productivity objectives; then rank and weight the factors; then, using the factors, calculate an index for each machine that identifies the machines more critical to the objectives.

An advantage of the method is that it can be molded to any plant's environment by adding or deleting factors and assigning different weights. Typically, the method goes through a period of fine-tuning in which weights are adjusted to better reflect the true environment.

At the Engine Works, a group of managers and supervisors responsible not only for maintenance activities, but also for total factory operations, ranked the factors from most to least important in meeting Maintenance Planning objectives. Such a broad-based group reflects the concerns of the many factory disciplines. Group-selected and ranked factors include (highest to lowest rank): dimensional tolerance capability; likelihood of breakdown; deterioration of machine and process without preventive maintenance; maintenance history of machine; danger of machine failure; spare-parts availability; ease of repair; and time since last overhaul.

The group also weighted each factor on a 0-to-100 scale. The higher the weight, the more important the factor in achieving objectives. The chart shows results of the Engine Works weighting. A plant making a different product with different labor, equipment, budget, priorities, and policies might come up with different ranks and weights.

The group also identified, ranked from most to least important, and weighted factors affecting Productivity objectives. (See chart.) Factors include: average projected machine load; number of alternate machines; transfer-line vs free-standing equipment; normal in-process inventory; average repair time; number of operators idled directly because of breakdown of the machine; and investment in the machine.

Every machine in the plant was then evaluated (scored) compared to the 15 factors identified for the Productivity and Maintenance Planning objectives. The group developed scoring tables for each factor and forms to record the scores. The score for each factor is between 1 and 10, the higher scores representing more important factors. Most factor scores are

---

Put the maintenance at the right place in the right amount at the right time doing the right thing—that's what a capital-intensive plant must do to make its valuable assets pay off.

Tolerance range, in.	Score
Less than ±0.0002	10
±0.0002 to less than ±0.001	9
±0.001 to less than ±0.003	7
±0.003 to less than ±0.010	4
±0.010 to less than ±0.030	3
More than ±0.030 or no tolerance	1

constant from year to year. Hence, once they are collected, most set-up work is complete.

To illustrate use of factor scores, the chart shows the development of scores for the top-ranked factor (tolerance) for Maintenance Planning objectives, and the top-ranked factor (projected machine load) for Productivity objectives. The chart is for five sample machines in a much larger total population really evaluated.

Machines with closer tolerances need more attention than others. Thus, the group developed a scoring system such that higher scores go to machines with closer tolerances.

An often-used machine should have more maintenance than one seldom used. Therefore, the group developed a scoring system such that higher scores go to machines

MAINTENANCE PLANNING INDEX

Rank, factor, and weight Machine	1 Tolerance capacity Wt = 100	2 Likelihood of breakdown Wt = 90	3 Deterioration without preventive maintenance Wt = 85	4 Maintenance history Wt = 75	5 Danger of failure Wt = 65	6 Spare-parts availability Wt = 60	7 Ease of repair Wt = 35	8 Time since last overhaul Wt = 20	Maintenance Planning Index, MPI
5655, horizontal knee type milling machine	7	2	3	3	1	7	1	1	1,900/530 = 3.58
3987, precision boring machine	7	2	3	1	1	10	6	10	2,285/530 = 4.31
4499, transfer line, final bore	9	5	7	2	10	7	3	1	3,290/530 = 6.21
9399, tapper	1	2	3	1	1	3	1	10	1,090/530 = 2.06
8766, cam grinder	7	2	3	10	8	7	3	10	3,130/530 = 5.91

PRODUCTIVITY INDEX

Rank, factor, and weight Machine	1 Average projected machine load Wt = 100	2 Number of alternate machines Wt = 90	3 Transfer line or free standing Wt = 80	4 Normal in-process inventory Wt = 60	5 Average repair time Wt = 60	6 Number of operators idled due to breakdown Wt = 40	7 Investment in machine Wt = 30	Productivity Index, PI
5655, horizontal knee type milling machine	5	5	2	1	1	4	4	1,510/460 = 3.28
3987, precision boring machine	5	5	2	5	4	1	4	1,810/460 = 3.93
4499, transfer line, final bore	10	10	8	7	7	7	6	3,840/460 = 8.35
9399, tapper	5	4	2	3	1	3	1	1,410/460 = 3.07
8766, cam grinder	5	10	2	2	4	6	1	2,190/460 = 4.76

Hours per day	Score
Over 24	10
16+ to 24	8
8+ to 16	5
Less than 8	1

expected to have greater use. Projected machine load data are usually readily available in most plants. Scores for the projected machine load factor would change if projected demand changes. Such factors should be re-evaluated periodically to assure that they reflect current conditions.

The Engine Works uses a wage-incentive system based on output, but machine load is measured in "standard" hours. Therefore, in 1 day of 3 consecutive shifts, the machine could be used for more than 24 standard hours. Thus, the score of 10 for machines with projected use of more than 24 hours per day.

The MCI

To simplify comparisons, scores for different factors for a given machine are combined into one number, called the Machine Criticality Index (MCI). It summarizes the importance of each machine to Productivity and Maintenance Planning objectives.

The summarization is in three steps that extract two additional indexes, which then combine to form the MCI. The additional indexes are the Maintenance Planning Index (MPI) and the Productivity Index (PI). For a given machine, each will be a number between 1 and 10, with 10 representing the most important or critical to overall Productivity and Maintenance Planning objectives.

Referring to the chart, to calculate the MPI for Machine 5655, for example, multiply factor scores by factor weights for that machine. The sum of such products for the 8 factors for Machine 5655 is 1,900.

Divide the sum by the total weights of 8 factors, 530, to find the weighted average of the scores, in this case, 3.58. This is the MPI. Similar calculations for the Productivity Index for Machine 5655 result in PI= 3.28.

Though the two indexes by themselves are sometimes useful, in most decision-making situations you need an overall index that accounts for your plant's interpretation of the relative importance of Productivity and Maintenance Planning goals; that is, you need the MCI.

The group at the Engine Works viewed Productivity objectives as more important and assigned the PI a weight of 100. Maintenance Planning objectives received a weight of 65. The chart shows the equation for determining the MCI from the weighted averages of the

$$MCI = \frac{100\,PI + 65\,MPI}{165}$$

Machine		MCI	Rank According to MCI
5655	milling machine	3.40	4
3987	boring machine	4.08	3
4499	transfer line, final bore	7.51	1
9399	tapper	2.67	5
8766	cam grinder	5.21	2

Payoff and proof

Here are some rewards that the John Deere Engine Works gained from the weighted-average method of maintenance planning:

Lower maintenance expense

■ Maintenance expense on the top 10 machines on the MCI ranking was reduced by 18.5% from 1978 to 1979.
■ Maintenance expense on the highest-ranked machine on the MCI ranking, a section of a high-volume, 6-cylinder diesel block transfer line, was reduced by 66.3% from 1978 to 1979.
■ Overall maintenance costs were reduced by $800,000 from 1978 to 1979, and the weighted-average method contributed significantly.

Lower scrap expense

■ On the head transfer line, a department consisting of machines with high MCIS, scrap as measured in dollars was reduced 20.2% in 1979.

Higher productivity

■ The 6-cylinder diesel block transfer line (the highest ranked machine on the MCI ranking) experienced a 5.7% improvement in time-on-incentive in 1979.
■ Utilization of the cylinder liner boring machine increased 9% from 1978 to 1979. This is the 15th ranked machine on the MCI ranking.

PI and the MPI.

The objectives could be weighted and ranked differently to reflect a different set of operating conditions and corporate priorities and objectives. This flexibility makes the method useful in many environments.

The calculated MCIs for the five sample machines in the chart shows that Machine 4499 is by far the most important machine in the sample group; Machine 9399, least important. From that informa-tion, a manager can decide how to allocate maintenance activities.

In use at The Works

At John Deere, computations are handled by a computer program which calculates indexes and ranks machines according to each of the three indexes. The ranking is updated once a year. In an environment where demands and product mix are less certain, updating should be more frequent, possibly as often as demand projections change for master scheduling.

The Engine Works and the Tractor Works use the three indexes as basic criteria for these short and long-range operating decisions:
- Daily maintenance priorities.
- Scheduling of maintenance overtime.
- Machines to include in the preventive maintenance program.
- Criticality of spare-parts inventory.
- Priority of special training for skilled trades.
- Machine replacement analysis, justification, and scheduling.
- Evaluation and selection of new machine tools.

There are also intangible benefits from this method. Many times, it helps to clarify or quantify a problem and alternate solutions. John Deere managers have found that it greatly enhances their knowledge and awareness—they can analyze and solve existing or potential problems more effectively.

The method gives consistent and unbiased figures on which to rely in making many decisions. It does *not* tell you what maintenance should be carried out, what should be an appropriate maintenance budget, how many people the maintenance department should employ, or any other action. It makes no decision for you. It processes information objectively so *you* can make informed decisions. The success of the method depends on its user.

Weighted average method lets Engine Works assign maintenance like servicing this coolant transfer pump for a crankshaft oil hole drill on the basis of machine criticality to overall productivity.

Technician with vibration analyzer and recorder checks head drives and screw feeds of cylinder head transfer line. Machine Criticality Index points out which machines bear closest watching for signs of degradation that could eventually cause surprise downtime.

144

Reprinted courtesy of Paul Juul & Associates Inc., Bellevue, Washington

Maintenance Management Surveys

This in-depth survey is based on data collected at over 150 Maintenance Management Seminars attended by over 4,000 plant engineers and maintenance supervisors representing 2,500 organizations.

The results of 8 years data collection and analysis are presented below:

Maintenance Performance Record: Less than 4 Hours Work Per day - Work sampling indicates that less than 4 hours a day is spent in hand-on maintenance work. The balance of the day is lost because of:

- Waiting for assignments
- Looking for supervisor to get next assignment
- Visiting the jobsite to see what must be done
- Making unnecessary trips to stores for tools and materials
- Waiting for workers of other crafts to finish their jobs
- Using two men when one would be enough
- Waiting for production to release equipment
- Searching for sufficient information or blueprints
- Taking extended coffee breaks, washing up early, starting late
- Conducting union business

Maintenance Management Organized at All Levels - Good maintenance management can be achieved only with good organization and a clear understanding of the responsibility and authority of all management levels.

1. Repair, alterations, or new construction - should be performed by foremen. A foreman should spend at least 6 hours with his men and no more than 2 hours a day in meetings or doing paperwork.

2. Clerical functions, filing, identification of materials requirements, work-order processing, scheduling and timekeeping - should be done by planners and maintenance clerks. Six hours should be spent in the office and a maximum of 2 hours a day in the plant. One planner should be able to plan the work of up to 30 mechanics.

3. Analyzing causes for failures, upgrading equipment, and designing preventive maintenance programs - should be the responsibility of the maintenance engineering function. One maintenance engineer can support a 100-man maintenance department.

These three functions can be performed by one person in small maintenance organizations, but they should be separated in large organizations.

. 1 percent to the owner of the business

. 45 percent to plant managers

. 14 percent to engineering functions

. 18 percent to operating superintendents (production)

. 14 percent to administrators or directors

. 8 percent to other positions

Almost 80 percent of all maintenance departments have up-to-date organizational charts and job descriptions for all key positions. And, most maintenance departments have communication problems with top mangement and the operating departments.

Forty-two percent have planners, 73 percent have full-time maintenance clerks, 28 percent have one full-time clerk, 33 percent have two to four clerks, and 12 percent have five or more clerks. All companies with five or more maintenance clerks also reported over 100 maintenance workers. Majority Dissatisfied with Work-Order System - Many companies have poor control over maintenance costs and performance because they lack a good work-order system.

The maintenance requisition, or work order, is the basic document used for authorization of maintenance work. It is an important file document that gives maintenance management the history of equipment failures, causes for failures, cost of repairs, and action taken to reduce failure rates. The work-order file is one of the main sources of information for plant engineering or maintenance engineering departments in designing new facilities or planning plant expansions.

Eighty-six percent of the maintenance organizations in the survey have formal work-request or work-order systems, and 61 percent of these companies use the work order for requesting and controlling all work. Thirty-nine percent use formal work orders only for major work.

Operations, production, and other departments served by maintenance request alterations or expansion of the facilities frequently, but, in most organizations, they do not share in the expense. Operating departments should be given budget responsibility for maintenance of the equipment and facilities they use. The owner of an automobile - not the garage fixing the car - must pay for repairs. It should be that way in industry, but the survey gave a very different picture.

Anybody can request work in 22 percent of the companies, but only operating department heads can make requests in 30 percent of the companies. Operating supervisors must request work in 38 percent of the companies, and engineering or planning personnel must initiate requests in 10 percent of the companies.

Persons authorized to approve maintenance work are most often maintenance supervisors. The survey showed that work orders must be approved by:

Plant engineer	14 percent
Maintenance supervisors	46 percent
Works manager, general manager	17 percent
Operating supervisors	17 percent
Any supervisor	6 percent

The work order provides communication between operating departments and maintenance. It also provides essential cost information, causes and frequencies of failures, and scheduling data. The survey revealed that:

. 52 percent of the companies use fewer than 100 work orders per week. The average work order was estimated to require about 8 man-hours of work.

. 60 percent of all companies use some type of priority coding

. 32 percent analyze causes of failures

. 28 percent break work order estimates into man-hours per craft

. 30 percent compare actual times to estimated or planned times

. 28 percent use computer support for the maintenance program

- 34 percent report weekly on the backlog of work orders

- 27 percent use backlog trend data for increasing or decreasing the work force

- 39 percent use general backlog data for scheduling overtime and subcontracting

- 35 percent express the backlog in man-hours per craft

- 20 percent have less than 2 weeks' backlog

- 26 percent have between 2 and 8 weeks' backlog

- 8 percent have between 8 and 16 weeks' backlog

- 4 percent have between 4 and 12 months' backlog

- 3 percent have over 1 year's backlog

- 39 percent of the responding maintenance supervisors did not know the status of their backlog

- 60 percent of the companies identify and secure parts and maintenance materials before the work order is released to the workers.

Forty-six percent of the participating plant engineers and maintenance supervisors were satisfied with their work-order systems. Or, 54 percent felt their work-order systems needed major improvements.

What Companies Are Doing About Maintenance Planning - Maintenance planning varies greatly from plant to plant, from the simple method to the complex computerized

system involving weekly work programs, preventive maintenance, and major PERT
(Program Evaluation and Review Technique) or CPM (Critical Path Method) planning
project.

The planning process should include material and parts identification, tooling,
make or buy decisions, estimating, assigning priorities, and scheduling. Planning
should be based on the principle that maintenance serves the plant's operations.
Maintenance work must be coordinated with the departments served. The survey
showed the following methods of coordination:

. 20 percent of companies have daily meetings between operations and maintenance

. 10 percent have weekly meetings

. 24 percent rely on maintenance supervisors for all coordination

. 6 percent use engineering for coordination

. 32 percent use planners or schedulers for coordination

. 8 percent rely on close personal supervision by firstline supervisors

Most repair jobs must be completed in a fairly limited time to prevent production
downtime. Jobs must be well planned and based on good time estimates. Very
few maintenance organizations base estimates on Work Measurement. Only 12 percent
use techniques such as work sampling, time study, MTM (Methods-Time Measurement),
or UMS (Universal Maintenance Standards).

. 20 percent of the companies have developed labor time standards for repetitive

jobs

- 46 percent use foreman's estimates for most jobs

- 25 percent have some type of productivity report comparing actual performance to planned performance

- 66 percent try to plan one or more days ahead

About 25 percent of the organizations use PERT or CPM for the planning, scheduling, and control of maintenance construction work.

Only 22 Percent Satisfied with PM Results - Most companies have a head start in their preventive maintenance programs, but only 22 percent of the participants are completely satisfied with results. Eleven percent are partly satisfied, and 67 percent are dissatisfied.

A preventive maintenance program requires total commitment. It must be an integrated system consisting of specific inspections at specified frequencies followed by required repairs. A PM program must be supported by a good planning, scheduling, and dispatching system.

The survey indicated that many preventive maintenance programs were poorly conceived, installed, and controlled:

- 72 percent of the participating companies had assigned specific equipment numbers to all key equipment

- 65 percent had identified critical equipment based on three consequences of failure: endangering safety or personnel, stopping production, and causing high repair costs

- 62 percent keep history cards on all major equipment, listing date of repairs. Only 35 percent record labor and material costs for all repairs made to such equipment

- 32 percent have specific repair and maintenance budgets for key equipment

- 70 percent use standing work-order numbers for preventive maintenance

- 69 percent use some type of diagnostic PM inspection

- 82 percent have established lubrication routes and procedures

- 63 percent have established PM inspection routes and procedures

- 38 percent have detailed PM inspection check sheets outlining what to inspect

- 64 percent have established tickler files or tubfiles, or generate PM work orders by computer

- 40 percent have automatic follow-up programs to ensure prompt repairs of substandard equipment identified by PM inspectors

- 73 percent have difficulty coordinating the preventive maintenance inspections with production

Maintenance Usually Controls Spare Parts - Maintenance material management is primarily concerned with holding annual inventory cost to a minimum while maintaining a satisfactory level of service to operations.

Proper control of maintenance spare parts, insurance items, materials, and supplies is important in reducing maintenance costs. Material costs range between 20 and 70 percent of the total maintenance cost. The average cost of carrying the maintenance inventory is estimated at 2 percent a month, or almost 25 percent a year.

Most maintenance departments have full control of (or responsibility for) this major cost factor. The survey revealed that maintenance material and spare parts were controlled as follows:

Maintenance department	54 percent
Purchasing	18 percent
General stores	11 percent
Accounting or administration	5 percent
Engineering	4 percent
Top management	3 percent
Computer section	3 percent
Other	2 percent

In 1971-72 survey, the average total inventory level was about 12 months - that is, total value of the inventory of spare parts and general maintenance materials was equivalent to one year's usage. This inventory level increased sharply during 1973-74, reaching a peak of 21 months. The 1976 survey indicated an average level of 10.33 months.

Only 32 percent of the participating companies use an "ABC" analysis to separate the 20 percent of the inventory items that account for about 80 percent of the annual needs (A items). Only 25 percent of the companies use sealed-stock two-bin systems for low cost items (C items, or the 50 percent of the inventory items that account for only 3 to 5 percent of the annual cost).

153

. 55 percent of the companies use Min/Max or calculated EOQ (Economic Order
Quantities) to determine order quantities

. 9 percent use computers to establish order points and order quantities

. 9 percent have informal systems based on experience of the storekeeper and
maintenance supervisors

. 13 percent reorder when an item is out of stock

. 14 percent use the "guess" method of reordering

. 41 percent tie the stores requisitions to the work order number

. 74 percent delete obsolete materials regularly

Most companies indicated a need for better control of reconditioned and original
spares that are charged out to equipment or to other departments but that are not
installed or used.

SUMMARY

Maintenance management is a major challenge to top management. The survey
indicates that basic objectives are not being met. Major improvements must be made
in organizational structure, work-order systems, estimating methods, planning
and scheduling procedures, preventive maintenance, and material management.

CHAPTER 4
COMPUTERS IN MAINTENANCE MANAGEMENT

Presented at the CASA/SME Autofact 5 Conference, November 1983

Micro-Computer Aided Maintenance Management System

By Kishan Shyamlal Bagadia
Johnson Controls, Inc.

The micro-computer has already invaded almost every aspect of industrial activity with varying degree of success. Now it is taking its first steps into attacking the problems of Maintenance Management. In today's highly advanced technological world, after automating a factory , management's next frontier for labor cost reduction lies in the maintenance & craft areas.

A basic maintenance management system consists of:

1. Equipment History

2. Preventive Maintenance

3. Workorder System

4. Spare parts inventory control

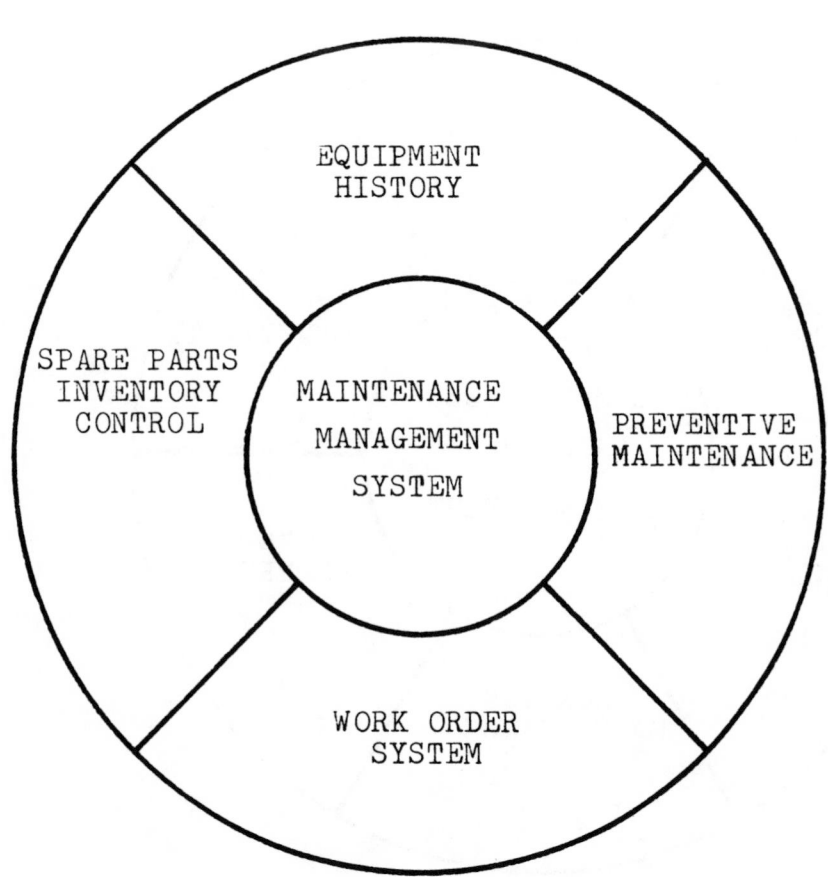

At present, most maintenance management systems are available on Mainframe computers only. With the recent drop in micro-computer prices, it is now possible to dedicate a micro-computer system for maintenance department.

A maintenance management system has been developed on micro-computers. A micro-computer will keep your maintenance records up to date. It will provide information with which you can decide if a given piece of equipment is being maintained at the best intervals. It produces a work order so the right man knows it is time to check out a specific piece of equipment and follow up to make sure he does.

Many small - and medium sized - companies will find this program to be well worth in terms of lower total production costs. For example, in assembly line operations, where a line shut down may mean a loss of several thousand dollars, an effective computerized maintenance system is an insurance against production stoppages and prohibitive maintenance charges. Similarly, a computerized maintenance system in the oil and chemical industries is an invaluable means of improving plant safety and operational efficiency.

A description of the program will readily show how a maintenance department can be effectively managed.

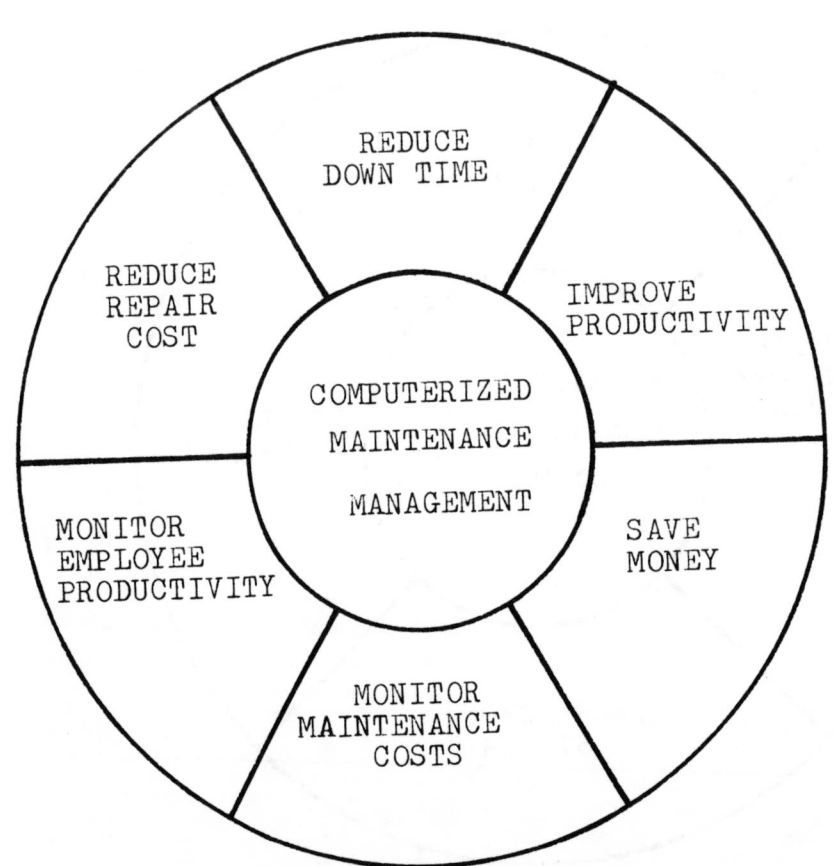

1.0 Equipment History:

Equipment records can be stored into the computer. The following information can be stored for each machine:

a. Manufacturer
b. Model
c. Serial Number
d. Purchase Date
e. Cost
f. Date Installed
g. Location
h. Description, etc.

Any of this information is readily available to the user by punching a few keys on the computer.

2.0 Preventive Maintenance:

Machine design techniques have become very complicated within the last few years. Regardless of what technological advances are made, there is one factor that must accompany all workable machines - RELIABILITY. A very effective and organized Preventive Maintenance (PM) program is required to insure reliability of machine and equipmentoperation.

PM consists of periodic inspection or checking of existing facilities to uncover conditions leading to production breakdowns or harmful depreciation and the correction of these conditions while they are still in the minor stage. The importance of an effective PM program cannot be emphasized enough. A significant amount of money can be saved with the help of an effective PM program. With the help of this software, PM instructions are stored by each machine number. The functions are separated by each craftsman, for example, electrician, pipe fitter, etc.. The user has the flexibility to design his own PM program and implement it.

The computer program allows the user to build a Master PM schedule which contains following information pertaining to each machine:

a. Equipment Number
b. Job Number
c. Craft Code
d. Description of PM Function
e. Frequency
f. Intended Starting Date

Based on these frequencies and intended starting date, the program displays the PM functions that have to be performed during a particular week. A computer output can be obtained so that the Maintenance Supervisor can issue Work Orders. (See Figure 1 for a sample output). A particular week could be accessed by entering 'Today's date'. After the PM function is performed the user enters the date of performance which is stored permanently.

The program also display PM activities performed on a particular machine up to date. (See Figure 2 for a sample output). This is a very valuable tool for maintenance manager. If an equipment is constantly breaking down, he can request a status chart for that equipment which would readily show if the PM functions are being performed or not.

The program also allows additions of new machines and PM functions. Obsolete machines can be deleted.

3.0 Work-order System:

A workorder system is used to satisfy the following purposes:

a. To provide authorization for the expenditure of labor and material in performance of the work.
b. To furnish a document for recording that the operation was performed.
c. To provide a document for the written feed back of other information such as material used, other work which was or may be required, etc..

The software prepares new workorders to authorize PM and other maintenance tasks. It accepts entry of labor grades and rates. It automatically generates a ten digit workorder number.

After completion important information such as, actual work performed, time spent, material, and labor cost, etc. is permanently stored in a file.

4.0 Spare Parts Inventory Control System:

Any maintenance department is required to keep a supply of materials and parts on hand if, it is to operate efficiently. Items that are continuously used such as; nuts, bolts, screws, nails, and so on are generally referred to as standard stock items. For these items a minimum quantity is established generally referred to as 'Re-order-point'. Items that are not part of the standard stock inventory are referred to as special purchase items obtained on the basis of individual procurement requests.

The software stores and retrieves information such as, part number, quantity, vendor, date, re-order-point, etc.. It updates quantity on hand based on transactions.

5.0 Other Reports:

5.1 Equipment Cost Summary:

Lists all units programmed into the system showing Equipment Number, Description, and Cost (See Figure 3).

5.2 Work-Order Charge Summary:

Work order charges (Labor, material, and total cost) for an individual work order or a group of work orders

(See Figure 4).

5.3 System Activity Report:

Shows the number of work orders printed and the number returned and processed for a given period.

5.4 Repair History:

Shows dates, description, and cost of repairs for an equipment.

Hardware Requirements:

64K Micro-computer
2 Disk Drives(Software is compatible with hard disk)
Monitor
Printer

Conclusion:

Plant management and engineering personnel have documents that identify the timing of all work within their areas and allows them to manage with a broader perspective. The use of this software can reduce the maintenance work force or cost. It helps maintain quality of the machines and thereby that of the products. Implementation takes a great deal of planning, co-ordination, and co-operation, but the results are well worth the effort.

```
          PREVENTIVE MAINTENANCE WEEKLY SCHEDULE   04/23/83


MACHINE#   JOB#    CRAFT    DESCRIPTION                      FREQ

445001      1      01       CHECK THE BATTERY                52   OVERDUE 2 WEEKS
445001      2      01       CHECK SPARK PLUGS                52
445001      3      11       CHECK OIL                        26   OVERDUE 1 WEEK
445002      1      03       CHECK HYDRAULIC PRESSURE         52
445002      3      09       INSPECT THE WELD AT LEGS         52
445002      4      07       CHECK THE INSULATION             26   OVERDUE I WEEK
445003      2      01       CHECK ALL SWITCHES               52
445003      3      01       CLEAN THE TERMINALS              52
T57001      2      01       CHECK THE WATER LEVEL            52
T57001      3      09       CHECK WELD   ALL MOUNTINGS       26   OVERDUE 1 WEEK
T57001      4      01       CHECK RELABY FOR WEAR            52
778001      2      11       LUBRICATE DRIVE MOTOR            52
778001      3      02       CHECK BELT & PULLEYS             52
```

FIGURE 1

PREVENTIVE MAINTENANCE STATUS CHART

STATUS REPORT PERIOD: 1/9/83 THROUGH 1/1/84

MACHINE #: 445001   JOB #: 4   CRAFT: 02   FREQ: 012

DESCRIPTION: LUBRICATE DRIVE MOTOR

```
- 1/9/83-      ---------      ---------      * 1/30/83*

---------      ---------      ---------      * 2/27/83*

---------      ---------      ---------      * 3/27/83*

---------      ---------      ---------      ---------

* 5/1/83*      ---------      ---------      ---------

*********      - 6/5/83-      ---------      ---------

*********      - 7/3/83-      ---------      ---------

---------      *********      ---------      ---------

---------      *********      ---------      ---------

---------      ---------      *********      ---------

---------      ---------      *********      ---------

---------      ---------      *********      ---------

---------      ---------      ---------      *********
```

FIGURE 2

02/27/83

EQUIPMENT COST SUMMARY

EQUIP NO.	DESCRIPTION	COST
019001	ASSEMBLY CART	4000.00
055001	GRINDER	200,250.00
055001	OH POWER & FREE	300,980.00
055002	FLOOR TOVEYER SYSTEM	1,200,000.00
055003	P & F CONVEYOR	34,000.00
055004	P & F CONVEYOR	45,000.00
441002	TURRET LATHE	123,789.00
445001	LATHE	100,000.00
445003	SCREW LATHE	200,450.00
445004	TURRET LATHE	250,980.00
445005	SCREW LATHE	230,000.00
601001	PAINT BOOTH	495,000.00
621003	PISTON PUMP	15,000.00
778001	PICK UP TRUCK	12,560.00
778002	PICK UP TRUCK	12,560.00

		$3,224,569.00

REPORT: EQUIPMENT COST SUMMARY

FIGURE 3

```
        02/27/82
     WORK ORDER COST SUMMARY                          PAGE 1

     WORK ORDER#        LAB COST      MAT COST       TOT COST
     ----------         --------      --------       --------
     441001001            59.50         37.50           97.00
     441001002            12.75          4.50           17.25
     055001001           132.60        257.00          389.60
     601001001           102.00         68.50          170.50
     778002001            76.50         36.50          113.00
     T57001001           340.00        259.00          599.00
     441002001            34.00         19.00           53.00
     621003001           170.00         27.50          197.50
     ----------------------------------------------------------
                         927.35        709.50        1,636.85
     ----------------------------------------------------------
```

WORK ORDER COST SUMMARY

FIGURE 4

Bibliography:

1. G. W. Allman and J. H. Bottom, "Scheduling maintenance
    projects with ease - Using a micro-computer", Plant
    Engineering, October 1980.

2. Kishan Bagadia, "Maintenance functions on micro-computers",
    1982 Fall Industrial Engineering Conference Proceedings.

3. Dr. C. L. Brisley and Royal J. Dossett, "Computer use and
    non-direct labor measurement will transform profession in
    the next decade", Industrial Engineering, August 1980.

4. Frank V. Claire, "Indirect labor measurement for profit",
    IMS, 38th. annual clinic, proceedings, November 1974.

5. John E. Heintzelman, "The complete handbook of maintenance
    management", Prentice-Hall, Inc, July 1977.

6. John E. Koop, "A simplified technique for pin pointing
    non-productive maintenance costs", Plant Engineering,
    April 1975.

Reprinted courtesy of Paul Juul & Associates, Inc., Bellevue, Washington

Problems in Maintenance Management

Conventional computerized manufacturing management and cost accounting packages, which work so well in production, don't work so well in the maintenance environment. In maintenance costs and schedules are difficult for estimate in advance. We don't always know how long any individual nonrepetitive maintenance job will take. We don't have the same bill of material every time we repair a given piece of equipment.

A solution is to estimate maintenance jobs on a statistical time slot basis. The statistical time slot estimates should be calculated on a 5 month moving average basis exponentially smoothed. By this means realistic averages are maintained automatically and adapted to the particular maintenance environment and level of skill of the maintenance estimators. Maintenance jobs have to be scheduled within the constraints of available manpower for each craft. In order to calculate the man-hour backlog for each craft, the system must also keep track of the average amount of over-time, emergency work, routine and preventive maintenance inspections, standing work orders, absenteeism, vacations and contract labor. This should be on a 5 weeks moving average basis. Maintenance scheduling has to be revised continually as priorities of the jobs change. Emergency and urgent jobs will always occur, and cannot be predicted. They frequently require immediate reassignment of craft and rescheduling of other work. Similarly, there is always a significant amount of routine and preventive maintenance inspection work done on standing work orders, which can be predicted statistically.

Studies have shown that the productivity (wrench time) of most maintenance employees is only 30 to 50% often through no fault of the men. Obsolete maintenance management practices are the main causes and make it difficult for employees to do their job effectively. Delays that prevent craftsmen from achieving results are frustrating and costly. Employees spend a significant part of their time waiting, or on non-productive tasks, such as:

- looking for foremen to get job assignment.

- visiting work site to find out what must be done.

- rounding up materials and multiple trips to the warehouse.

- return trips for tools.

- spare parts out of stock, no information on substitute parts.

- waiting for other crafts to finish.

- a shortage of crafts to finish parts of the job.

- waiting for a shut-down, clearances, and access to job site.

- waiting due to lack of information or drawings.

- losing time because of countermanded orders.

- rescheduled job priorities.

- extended coffee or lunch breaks

- late start up, early wash up

- waiting for special tools or engineering specs.

- assigning more men than required (Buddy system).

A comprehensive well-designed computerized maintenance management system cuts delays dramatically through systematic planning, scheduling, and control by:

- making materials and special tools available when needed, at the job site, <u>before</u> starting.
- providing clear instructions.
- co-ordinating work between the crafts.
- providing adequate drawings.
- agreement in job priorities.
- up to date history records.

The result will be a significant increase in productivity of the maintenance work force, reduction of maintenance costs, reduction of overtime and outside contracts. Frustrations will be reduced and morale improved. The maintenance backlog will shrink, equipment downtime will decrease, and overall plant productivity will rise.

The keystone of the maintenance organization is the foreman. He's the first-line leader of his crew. He assures that the organization's policies, programs and work schedules are executed in the prescribed ways. The foreman must ensure quality of work, issue instructions, co-ordinate jobs, and reduce idle time when progress is delayed. Foremen are too important to spend time writing requisitions, looking for parts, and doing time-keeping chores. The maintenance management system must reduce the foremen's paper work so they can spend more time supervising work and solving specific technical, as well as, "people" problems.

A comprehensive, well-designed computerized maintenance system makes this possible. Managers and supervisors are furnished timely information about work progress, backlog of work, how much work can be accomplished, what jobs are scheduled, what jobs are waiting and why, variances from budgets, etc. Managers are responsible for the success of the maintenance programs. They must implement practices that make it possible for the employees to be efficient, quality oriented and safety minded

A good computerized maintenance management system is an effective tool to obtain increased productivity through:

- uniform methods to initiate work.

- procedures to approve and authorize work.

- systematic planning and scheduling procedures.

- formal consistant procedures to estimate jobs.

- detailed procedures and established defect limits
  for preventive maintenance.

- equipment data and spare parts data base.

- weekly review of work priorities.

- weekly work program based on craft net capacity
  to do work.

- inventory information on maintenance materials and parts.

- uniform methods to measure and control work backlog.

- recording man hours, materials, and job costs.

- uniform procedures to report completed work.

- development of maintenance history and downtime records.

- information on equipment performance and failure analysis.

CONDITIONS FOR SUCCESSFUL IMPLEMENTATION:

Successful maintenance management implementation consists of three phases: First, there must be a general maintenance management review to identify improvements that may be needed in:

- organization.
- work order systems.
- backlog control.
- planning and scheduling procedures.
- materials management.
- preventive maintenance.
- failure analysis.
- computer hardware and software - capacity and priority.
- management reporting system.

The second phase is software development, computer systems implementation, and systems testing.

The third phase consists of the all important human engineering aspects of user implementation on the shop floor. This includes training of maintenance supervisors and key personnel.

The combined three-phase approach will guarantee that the computerized maintenance system will not just work well in the computer room, but will also work well on the shop floor and become an accepted every-day tool for maintenance personnel and management.

DATA BASE AND SUPPORTING MODULES

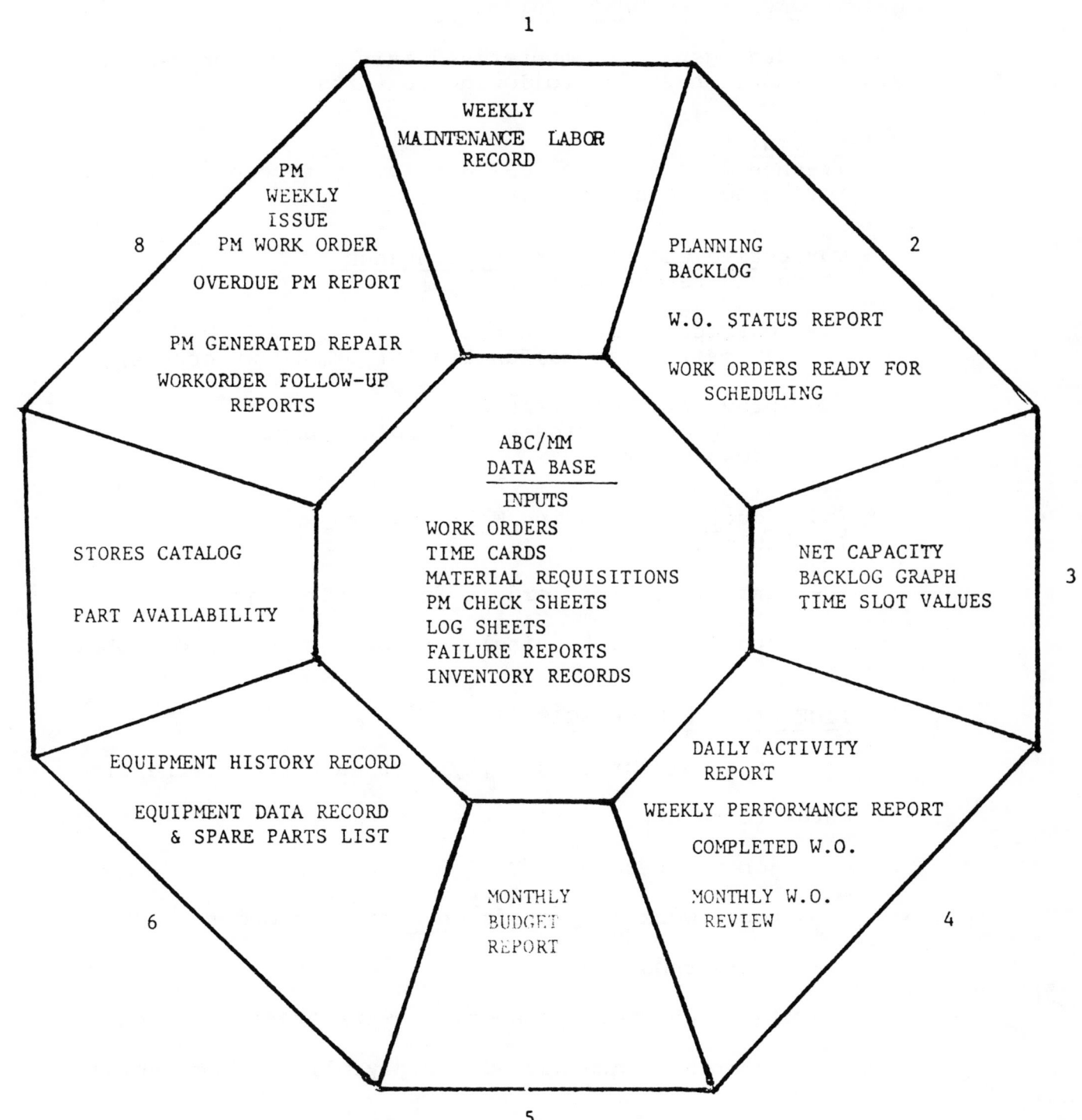

1

WEEKLY
MAINTENANCE LABOR
RECORD

PM
WEEKLY
ISSUE
PM WORK ORDER

OVERDUE PM REPORT

PM GENERATED REPAIR
WORKORDER FOLLOW-UP
REPORTS

8

PLANNING
BACKLOG

W.O. STATUS REPORT

WORK ORDERS READY FOR
SCHEDULING

2

ABC/MM
DATA BASE
─────────
INPUTS
WORK ORDERS
TIME CARDS
MATERIAL REQUISITIONS
PM CHECK SHEETS
LOG SHEETS
FAILURE REPORTS
INVENTORY RECORDS

STORES CATALOG

PART AVAILABILITY

7

NET CAPACITY
BACKLOG GRAPH
TIME SLOT VALUES

3

EQUIPMENT HISTORY RECORD

EQUIPMENT DATA RECORD
& SPARE PARTS LIST

6

DAILY ACTIVITY
REPORT

WEEKLY PERFORMANCE REPORT

COMPLETED W.O.

MONTHLY W.O.
REVIEW

4

MONTHLY
BUDGET
REPORT

5

Reprinted courtesy of Paul Juul & Associates Inc., Bellevue, Washington

Computer Software and Hardware

A. FEATURES OF THE IDEAL COMPUTERISED
MAINTENANCE MANAGEMENT SYSTEM.

    An ideal general computerised maintenance management system should have the following features.

* Generality:
It should be able to deal with every type of fixed, or moving equipment or plant.

* Corrective and Preventive Maintenance.
It should handle both

    - corrective maintenance
(repair after malfunction or breakdown occurs)

    - preventive maintenance
(inspect, lubricate and replace before breakdown occurs)

* Individual and Standing Work Orders.
It should handle both

    - individual work orders for specific job.

    - standing (continuing) work orders for minor jobs.

1. TIME CARDS AND CHARGES.

    It should process daily, for each shift, the time cards to report

    - job progress

    - hours worked by craft on each work order

    - hours remaining

    - labor charges for specific work order

    - interface data needed for payroll and accounting

2. WORK ORDER PLANNING & ESTIMATES.

    It should sort and display

    - work orders by multiple priorities

- multiple crafts involved

- individual and standing work orders

It should determine the work order backlog for each craft.

Estimating: It should provide

- labor estimates

- number of man-hours by craft

- estimate labor and material cost, prior to authorization.

3. SCHEDULING.

It should schedule the work weekly according to the available net capacity of each craft, and match the manpower with the workload considering priority requirements of all operating departments.

4. STATUS.

It should provide data for planners to determine the need and availability of parts, drawings, special tools, clearances, engineering, etc., and keep track of work order status such as

- waiting for authorization

- waiting for parts

- waiting for engineering

- waiting for craft shortage

- waiting for shutdown

- waiting for scheduling

- in process

- completed.

5. MULTIPLE PRIORITIES.

It should simultaneously schedule and co-ordinate

- multiple priorities of work
  (e.g. emergency, critical, normal, waiting for
  shutdown)

6. MULTIPLE CRAFTS.

It should simultaneously handle

- multiple crafts
  (e.g. electrician, mechanic, carpenter, etc.)

7. JOB OR TASK STANDARDS.

It should provide for job standards detailing the work to be done for repetitive jobs, both for corrective and preventive maintenance.

8. AUTOMATIC JOB ESTIMATING STATISTICS.

It should provide data for estimating the average duration of different maintenance jobs, for each craft, using time-slot averages. The time-slot averages should be obtained by automatic historical analysis of actual time cards, for each craft and time-slot, on a moving average basis, preferably exponentially smoothed.

9. AUTOMATIC CRAFT CAPACITY STATISTICS.

It should provide data for estimating allowances, that must be made for average overtime, contract labor, unforeseen emergency and critical jobs, minor and standing work orders.

These averages should be obtained by historical analysis of actual time cards and work orders, on a moving average basis, preferably exponentially smoothed.

10. BACKLOG TRENDS.

It should provide a table or graph showing the backlog in weeks for each craft. This is needed for estimating the time between receipt and scheduled completion date for regular work orders.

Backlog graphs by craft are also an important management tool for determining whether an imbalance exists between different crafts, whether more or less contract labor is needed, and as a gauge of how well maintenance management is doing its job.

11. PERFORMANCE CONTROL.

Daily:
The foreman should get a daily acitivity report for his men showing

- how much time they worked

- on what work orders, during the previous shift.

This serves as the first line supervisor's daily control of his work force. It shows the original man-hours estimated, the actual hours worked per each craft, the amount of overtime, and the estimated remaining man-hours (backlog).

Weekly:
A weekly performance report is an important document needed for the weekly scheduling meeting. It highlights variances and "sleepers" i.e. forgotten work in process.

A weekly list of completed work orders is desirable. This informs planners and maintenance managements of variances from the original plan, insures upgrading of estimates, and explanation of major variances.

When a job is done, the foremen fill in the completed work order, the maintenance action taken, the date completed, primary and secondary cause code and sign it off, after approving the quality of the work.

The planner or clerk will change the work order status to "complete" on the screen.

Once a work order is completed and signed off, data should automatically be transferred to the equipment history file. Moreover, completed work orders should not be closed out immediately and deleted fromthe work order files. If purchased material is involved, the material charges are inevitably late, and should be included in the equipment history file.

Monthly:
A monthly work order review report is also desirable. This report gives the planner the actual total labor and material cost, and highlights the priority distribution, and the specific reason for doing the work (cause code). Primary and secondary cause codes are selected by the plant engineers to suit their analysis. They may be such items as

    (a)   normal wear
    (b)   abnormal wear
    (c)   faulty parts
    (d)   outside cause
    (e)   operator error
    (f)   overload
    (g)   acts of God
    (h)   safety
    (i)   environmental
    (j)   vandalism, etc.

The cause codes are kept in the equipment history file. They are useful to maintenance superintendents, plant engineers and others responsible for maintenance cost and equipment evaluation.

12. WORK ORDER ACCOUNTING.

The system should keep track of the daily labor, material and other costs charged to each work order. Note work order costs cannot be closed out as soon as the work order is completed, because some charges, such as parts invoices, may arrive months later.

13. EQUIPMENT DATA.

Equipment data cards, providing nameplate and technical details of each piece of equipment, spare parts numbers, vendor, substitute parts, critical measurements, etc. are needed by the planners and other maintenance personnel.

14. EQUIPMENT HISTORY AND DOWNTIME.

It should provide equipment history showing work orders, labor hours and repair costs, accumulative costs, breakdown causes, and downtime statistics for each piece of equipment.

The equipment history data and downtime data are needed to provide information to upgrade the equipment through failure analysis. It helps the planner by showing what repairs were previously made to the equipment, what are weak components, what were the work orders, labor and material involved. It also shows the amount of downtime

for each piece of equipment. In some cases, such as
utilities, downtime statistics are needed for outage
reports to various official bodies. Downtime statistics
are also valuable for justifying changes in maintenance
budgets, where decreased downtime leads to increased
profit from production facilities.

15. <u>MAINTENANCE BUDGET CONTROL</u>:

A maintenance budget control report is desirable for
each major cost centre or department. It should show the
originating and authorizing department and the actual labor
and material charges during the month. Preferably, this
budget should be made out by the production department
requesting the maintenance.

The report compares the budget authorized by the given
cost centre, or production department, versus the work
performed and the charges by the maintenance division
for that production department. The distribution of
man-hours charged during the month, by priority, and is
important indicating whether an excessive amount of
"emergency" and "critical" maintenance is requested.

It also indicates whether an excessive amount is
charged to "standing" work orders.

The amount of work performed by the maintenance
department for a given production department is really
controlled by requests from the production department.

By making the production department responsible for

the maintenance budget, they become more conscious of the costs involved, and can relate these costs as a figure per unit of production. In most cases, increased maintenance will reduce downtime and increase production.

In some cases the production department needs data to relate the value of the increased production to justify appropriate increases in the maintenance budget for the department.

This gives top management tangible information on the relation between better maintenance, less downtime, and increased production. If the maintenance department alone is responsible for setting the maintenance budget, they may find it difficult to justify increases, not knowing the production needs. Also, if the maintenance department sets the maintenance budget, they have little control if some production departments overshoot the budget by demanding excessive amounts of overtime or critical maintenance.

16. MAINTENANCE MATERIAL INVENTORY CONTROL.

It should provide a stores catalogue showing spare parts and substitute parts carried in stock. On line CRT screens for inventory inquiries are desirable to show, for each part specified, numbers on hand, number reserved, on order, order point, order quantity, vendor name, substitute part number.

A periodic "where used" list for stock items is

desirable. Otherwise, if a piece of equipment and all its spare parts are scrapped, some of these spares (eg bearings) may turn out to be needed for other equipment.

Material requisitions for maintenance work orders should be costed and charged against work orders, and deducted from inventory.

The planners should be able to check that all key parts are available for a work order, and reserve them, usually several weeks before the work order is released to the shop. Minor parts and consumables are generally handled by spot requisitions, when the work order is issued to the maintenance men.

When the status of a specific work order is "waiting for parts", whenever these specific parts are received by stores, automatic "back order fill" and "parts received" notices for the planners are desirable, so they can change the work order status and release the work orders for scheduling.

For inventory management, periodic reports on costs, type, and number of parts used in the period, turnover, and the number of days supply in stock are desirable.

Inventory cost reports and physical count reconciliations are desirable.

Apart from consumable parts, it is difficult to forecast the needs for maintenance parts, the demand for which is very irregular. Sometimes critical parts may not be used for years, but must be kept in stock because they cannot be quickly obtained from the vendor. Sophisticated statistical methods of forecasting are therefore often misleading and seldom useful for maintenance parts (unlike for parts used in regular production) except for consumables.

17. PURCHASING INFORMATION.

Daily shortage lists are needed for purchasing, showing all parts that have reached order point, or stock out. Backlog lists are also desirable showing those parts where the number reserved exceeds number of hand, plus number on order.

18. PREVENTIVE MAINTENANCE STANDARDS.

For each piece of equipment the preventive maintenance tasks that have to be performed, and their frequency (month, weeks, days, hours) or usage (days, hours, miles, kilometers) should be specified on the PM worksheet. Each piece of equipment should be able to have any number of tasks, with different frequencies (e.g. task 1 - once a week; task 2 - every 3 months; task 3 - every 12 months). The preventive maintenance system should issue PM work orders automatically on a weekly, or other pre-specified basis.

19. PM Follow-up.

Preventive Maintenance system loses its value if PM

generated work orders are not properly followed up, and needed repairs are not promptly made.

The PM follow-up system should issue weekly reports of overdue PM inspections. It should also follow-up corrective work orders generated as a result of earlier PM inspection. The corrective work orders should be handled by the work order planning and scheduling system previously described, but should be separately flagged and followed up within the preventive maintenance system.

20.   ON-LINE FEATURES.

Input:
It is desirable that the input is done on-line on a CRT screen, and that editing and correction of errors are performed on-line. The clerk should choose the input form by means of a "help" menu. An example is shown on Page 114, which shows the menu, and Page 87, which shows a specific input form. If the clerk makes a mistake, such as entering alphabetic data in the numeric field, the program should immediately flag the error and allow him to re-enter it. At the end of the screen, it is desirable to have the computer ask if all the above items are correct. This gives the clerk a last chance to scan all items for correctness, before it goes to further processing. If the clerk says "Yes", it will go to the update sub-program. If the clerk says "No", he can go down the screen again and will have a chance to correct the mistake he spotted.

A clerk or planner in the maintenance department enters data on-line all day on a screen. The data may come from time cards, requisitions, new work orders, job estimates for work orders, or changes in work order status.

Inquiry:
On-line CRT screens are desirable for inquiries by planners, and occasionally maintenance men and foremen, for the following:

(a)   Equipment, nameplate and history data for specific pieces of equipment. Frequently maintenance men also want hard copy of what is shown on the screen, so they can send for a special part, drawing numbers, tools, etc. which are indicated.

The screen should therefore have an on-line teletype or printer.

The printer can also be used as an audit trail during first implementation, when there are always arguments about what was entered, and charges that"the computer didn't do it right".   The only way to trace these problems is to have an audit trail, which also serves as hard copy back-up.

(b)   When working up a specific work order, planners need equipment data, and also parts availability from the inventory system. The inventory status should be shown on-line, including whether a specific part is on-hand, reserved, or on order, what are substitute parts, who is the vendor.

The cost should be shown, so the planner can see costs to date or charge or enter a material cost estimate.

The planner also writes out a stores requisition, and if necessary makes out a purchase order, for the key parts for the work order. The stores requisition is attached to the work order. It is later released to the maintenance foreman so his craftsmen don't have to spend time making out requisitions.

When making inventory status inquiries, planners or foremen also frequently want a hard copy of what is on the screen to discuss details with a foreman, purchasing agent, or engineer.

CRT Screen Width:
80-character-wide screens are adequate for input, but for inquiry 132-wide screens are preferable and save time, as more information can be displayed without scrolling.

Data Currency and Updating:
It makes little difference to the currency of the maintenance management information that you can see on the CRT screen, whether updating of the database is done instantly on-line, or in batches once per shift.

There are many reasons for this. For example, time cards are usually handed in at the end of each shift. Many planners, most foremen and other maintenance personnel usually originate data or make changes on paper forms, which are subsequently entered on a screen by a data entry clerk.

One possible exception is inventory, if each maintenance store clerk is provided with a computer terminal to immediately record issues and receipts. However, in most cases, the stores clerks collect but don't enter all the parts requisitions, or tear-off tags attached to the parts, and then send them over to a data entry clerk, who then enters this data in batches. The number of data entry clerks is usually minimized to save cost, so they can deal with an <u>average</u> but not a peak work load. Since the paper comes in batches, and since the time cards are from the last shift anyway, most of the information in the system will therefore be current up to the last shift only.

From the data processing viewpoint, the immediate updating of the data base, as soon as the data is entered on line on a screen, may be ideal. As explained, the currency of the maintenance information, however, will not be greatly affected if the data is entered on line during the shift, and the database is updated in batch, at the end of the shift or day. Either method produces about the same end result in the maintenance environment.

B.   MAKE OR BUY SOFTWARE.

   1.   GENERAL:

It is generally agreed that it costs about 10 times as much, and takes several years longer, to develop your own computerised maintenance management program, than to buy an existing maintenance management software package.

However, sometimes the maintenance needs of a company may be so specialised that no suitable package can be found.

Relatively few organisations have developed their own customised maintenance management program. Those that have done so, usually took several years, at a cost of several hundred thousand dollars. They had to assemble a team   of maintenance engineers, computer systems and programming experts, and perform analysis, developing, testing and debugging successive modifications of their system until their system worked well in the difficult maintenance environment. No one should under-estimate the magnitude of this task. One cannot adapt systems developed for other purposes.

   2.   SPECIAL MAINTENANCE REQUIREMENTS.

A successful maintenance management system differs radically from conventional production control, material requirements planning, or work order or cost accounting systems. One reason is that maintenance work orders are never as rigid and predictable as production work orders.

Only a minority of maintenance jobs are truly repetitive, such as preventive maintenance. The majority of individual maintenance jobs are non-repetitive and it is very difficult to estimate beforehand how long an individual maintenance work order will take. However, if groups of maintenance jobs are treated on a statistically averaged "time-slot" basis, it is possible to estimate fairly accurately how many man-hours they will take as a group.

Another difficulty is emergency jobs (e.g. breakdowns where life, property, or production lines are in danger) or critical jobs (e.g. clearing) which cannot be predicted, yet a successful maintenance management system must make substantial allowances for their inevitable occurrence. Another problem are the numerous minor maintenance jobs, which would take too much paper work, or be uneconomic to control individually, but which are numerous (e.g. changing burnt out light bulbs). Such minor maintenance jobs are usually handled by standing (continuing) work orders.

The scheduling of the maintenance work force by matching the craft capacity to a changing work backlog is much more difficult in maintenance than in production control. The maintenance workers are more highly skilled and diversified than production workers, have different crafts and specialisations, and are not necessarily interchangeable, even within one craft. (e.g. only 1 or 2 electricians may be trained in repairing a certain piece of electronic equipment). The result is that a maintenance management system must allow unlimited flexibility for continually rescheduling jobs of changing priorities using multiple crafts. Yet, it must provide a reliable estimate for what the total maintenance backlog, by craft, is each week, and how much of this work load can be performed during the week by each craft.

The foremen need to know the amount of work performed by each individual against schedule target each day. The maintenance superintendent must know the amount of work performed by each craft under the supervision of each foreman relative to the target.

At a weekly scheduling meeting the priority of all jobs must be reassessed in the light of constant changes. Maintenance work orders that have been released during the week must be added to the work load, allowance must be made for finishing jobs in progress, and allowance must also be made for unforeseen emergency and critical jobs. Then a new flexible schedule must be drawn up for the coming week.

The maintenance system must allow the prediction, on a statistical basis, of the man-hours available for each craft, after due allowance for vacation and absenteeism, average amount of overtime, contract labor, emergency and critical jobs, etc.

In short, the maintenance management system must provide the utmost flexibility in work order scheduling in a difficult, constantly changing maintenance environment.

3. MAINTENANCE INFORMATION FLOW.

Data processing people wishing to develop a maintenance management system must constantly keep in mind that a successful maintenance management system is primarily a "people system" supported by a paper flow system, assisted by a computer, and not the other way around.   There have been many failures caused by over-emphasis of the computer aspects and inadequate provision to allow for each and every one of the difficulties in the real life maintenance environment.

An ideal "paperless" purely on-line system, where everything is handled on CRT screens, is not practical in maintenance.

Maintenance information is best entered, queried and displayed on-line, but maintenance personnel and foreman want and must have the information they need on a form or piece of paper they can carry with them to the jobsite, and make notes on.   This paper may serve as a turnaround document, or it may lead to the issue of more paper and later on-line entries, after discussion with the foreman.   It is unrealistic to expect a maintenance man, or his foreman to carry a CRT terminal with him when he inspects and repairs a fan on the roof.

C.   SOFTWARE PACKAGE SELECTION.

1.   GENERAL.

A number of maintenance software packages will be listed in the next section. We obtained information on these packages through a careful literature search, correspondence and sales literature received, but it is based on the claims of each vendor, and we cannot guarantee its accuracy.

If one is planning on buying a maintenance package, one should carefully evaluate:

(a)   the functions of the package, including all input and output, and how it suits the company's maintenance requirements.

(b)   program language, data processing efficiency.

(c)   equipment requirements

(d)   documentation

(e)   provisions for maintenance and enhancements

(f)   vendor reputation

(g)   price

(h)   installation on computer

(i)   training services offered and cost

(j)   implementation cost on shop floor.

Apart from user training for data entry clerks and planners, it is desirable that training seminars are available at specific locations, or at your plant, for all maintenance supervisors and personnel, prior to start up. This is essential to ensure that the system will

really work well on the shop floor, instead of just in
the data processing department.

Beware of selecting a software package purely on
the basis of price, or data processing sophistication
or efficiency.

The software price and computer running cost will
usually be far less than your investment in training,
implementation and continued operation on the shop floor.

Top management's evaluation of a computerised maintenance management system is not its sophistication or
how efficiently it runs on a computer, but whether the
maintenance foremen are convinced it helps them, it works
well on the shop floor, and it actually cuts overtime and
contract labor, reduces backlog and downtime.

2. GENERALISED MAINTENANCE PACKAGES AVAILABLE.

Through a literature search, we have found only
five general maintenance management systems that claim
to deal with all types of fixed and moving equipment, and
satisfy the majority of the requirements listed previously.
These five general purpose maintenance management packages
are listed in Table I. It shows the main features of
these packages, the programming language, and computer
hardware, as indicated in their sales literature, supplemented in some cases by telephone inquiry. In
Table II, we give a list of the authors names and
addresses for these packages.

Supplier	ABC Mgmt.	Bonner & Moore	Decision Sciences	Mainstem Inc.	McDonnell Douglas
Name of Package	ABC /MM	Compass	FSMS	Mainstem On-line	PERMAC
FEATURES					
1. Time Cards & W.O. Charges	Y	Y	Y	Y	Y
2. W.O. Planning	Y	Y		Y	Y
3. W.O. Scheduling	Y			Y	Y
4. W.O. Status & Backlog	Y	Y	Y	Y	Y
5. Multiple Priorities	Y	Y		Y	Y
6. Multiple Crafts	Y	Y			Y
7. Job or Task & & Standards	Y		Y		Y
8. Statistical Job Standards	Y				
9. Statistical Craft Capacity Standard	Y				
10. Craft Capacity/ W.O. backlog balancing	Y			Y	
11. Performance & Control Reports	Y	Y	Y	Y	Y
12. W.O. Accounting	Y	Y	Y	Y	Y
13. Equipment data	Y	Y	Y	Y	Y
14. Equipment history data, downtime	Y	Y	Y	Y	Y
15. Maintenance budget control	Y	Y		Y	Y
16. Maintenance Parts inventory control	Y	Y	Y	Y	Y
17. Purchasing Information	Y	Y	Y	Y	Y
18. Preventive maintenance (pm)	Y	Y	Y	Y	Y
19. pm follow-up	Y	Y	Y		Y
20. On-line	Y	Y	Y	Y	Y
Computer Hardware	Any	IBM	Any	NCR	IBM
Program Language	Cobol	Cobol	Fortran	Cobol	Cobol

TABLE II – Suppliers of Generalised Maintenance
Management Computer Packages

1. ABC Management Systems Inc.
   Suite 3, 805 Dupont Street
   Bellingham, Wash. 98225   - Dr. V.W. Ruskin
                                    206-671-5170

   Equipment: Any major mainframe,
                    64K bytes 2 disks.
                    IBM, 34, 4300, 370, 30XX, DEC, HP3000,
                    DG, Univac, Honeywell, etc.
   Language :   ANSI COBOL

2. Bonner & Moore Inc.   - Mr. John W. Schatz
   2727 Allen Parkway   713-522-6800
   Houston, Texas 77019
   Equipment: IBM 8100, DG
   Language :   ANSI COBOL

3. Decision Sciences Corporation   - Mr. Hank M. Vasen
   Benjamin Fox Pavillion   215-887-1970
   Jenkintown, PA 10946
   EQuipment: Any
   Language :   FORTRAN

4. Mainstem Inc.   - George A. Prince, Jr.
                                  714-955-0362
   Note: Process customers' equipment received by mail.
           Also sell package with NCR computers
   EQuipment: NCR OEM
   Language :   ANS COBOL

5. McDonnell Douglas Automation Company   - Mr. W.R. Vickroy
   Box 516, Bldg. 107   314-232-8021
   St. Louis, MO 63166
   Equipment: IBM 370, 4300, 30XX MVS
   Language :   ANSI COBOL

3. <u>SPECIALISED MAINTENANCE PACKAGES AVAILABLE</u>.

Apart from the five general purpose maintenance management systems in Table I, we have found and listed in Table III seven specialised computer packages suitable for specific preventive, car, truck and aircraft maintenance. While they are designed to do a good job in their specialised field, they do not have all the features listed in Table I, which are desirable for a general purpose maintenance management system.

TABLE III – Suppliers of Specialised Vehicle, or
Preventive Computer Packages

1. Anawan Computer Services   - Mr. David G. Johnson
    18 Waterbury Lane
    Rehoboth, Mass. 02769
    617-252-4537
    Specialty   :   Preventive maintenance and
    Equipment   :   TRS80 - Basic

2. Marchbanks & Partners   - Mr. Gerald Parker
    4538 Centerview Drive
    San Antonio, Texas 78228
    512-736-1909
    Specialty   :   Vehicle
                    Fleet maintenance
    Equipment   :   Burroughs B1900   - Cobol

3. National Technical Information   - Mr. Edward Hicks
    5285 Park Royal Road
    Springfield, VA 22161
    703-557-4763
    Specialty   :   Vehicle equipment management information system.
    Equipment   :   IBM 370 OS DOS   - Cobol

4. Nepenthe Programs   - Mr. Will Hagenbusch
    3014 Briggs Court
    National City, CAL 92050
    714-475-5286
    Specialty   :   Aircraft Maintenance System
    Equipment   :   TRS 80 - Basic

5. Transportation Concepts & Services   - Mr. Lester A.
    20 Highland Avenue
    McLenden, N.J. 08840
    201-548-1200
    Specialty   :   Vehicle maintenance control system and
                    preventive maintenance service.
    Equipment   :   IBM OS VS DOS COBOL

6. USS Engineers & Consultants   - Mr. Frank Griffith
    600 Grant Street
    Pittsburgh, PA 15230
    412-391-8115
    Specialty   :   Preventive maintenance management
    Equipment   :   HP3000, Fortran

D.   COMPUTER HARDWARE REQUIREMENTS.

1.   COMPUTER.

The ideal maintenance management system should run on any general purpose mainframe or mini-computer, or even micro computer.

For ease of maintenance, and interfacing with the company's existing payroll, general ledger and other financial systems, the program should preferably be written in ANS 74 COBOL. That way they are also usually interchangeable between different makes of machines, such as IBM, Honeywell, Univac, DEC, HP, DG, or equivalent major makes. The programs should preferably run a minimum 64 K byte partition.

(a)   If a company has an existing mainframe with spare capacity, it is fastest, can use more CRT screens, and best suited for large maintenance staffs, plants and utilities. Mainframes also make it easy to interface with existing accounting programs like payroll and general ledger, and inventory. An on-line operating system is desirable. Most popular mainframes are IBM 34/370,4300/30XX, Honeywell, Univac, CDC, etc.

(b)   If a company already has a mini in the production department, or if a computer is dedicated to the maintenance function

Popular minis are IBM 34, 23, DEC, HP, DG, Wang, Honeywell 6.

Minis cost less than mainframes.

(c) For very small maintenance shops, a dedicated micro computer could be used. Its advantages are low cost. The disadvantages are less ease of operation, single CRT screen, slow speed, and limited capacity. Popular micros are Apple, Radio Shack, the new IBM personal computer, and many others. Micros could be used only in a small shop to provide a low cost "maintenance engine".

However, hard disks rather than floppy disks should be used, since floppy disks are easily damaged, and have insufficient storage.

(d) Time sharing and network services are advantageous in some cases, particularly when:

- there are multiple remote maintenance locations which cannot all have their own computer.

- if there are plants where it is not economically justified to hire computer personnel for three shifts per day operation.

- if there are plants in remote locations where turnover or data processing personnel shortage make in-house computers impractical.

2. DISK STORAGE REQUIREMENTS.

Large amounts of disk storage are usually required for equipment data, history, maintenance parts inventory, preventive maintenance files, and work order master files. The disk storage depends on,

- number of maintenance people
- number of backlog work orders
- number of inventory parts
- number of pieces of equipment
- how many years of equipment history have to be kept.

For example, a paper mill with 150 maintenance people requires about 75 megabyte storage.

On the other hand, a small maintenance shop, with only 25 people, 1000 inventory parts and 100 pieces of equipment, would require two 5 megabytes disks.

E.   PREPARATION FOR COMPUTERISATION.

(1)   Maintenance Audit conducted by task force
         comprised of maintenance and production
         supervisors supplemented, if desired, by
         consultants.

         Purpose is to identify maintenance require-
         ments the new computerised system should
         meet, changes that may be needed, areas that
         need improvement.

         Identify organisational and other changes
         needed (e.g. separate maintenance planners).

(2)   Draw up new system flow chart.
         Review by all maintenance supervisors.
         Discuss, modify and change until concensus
         obtained. It is very important to involve all
         maintenance supervisors at this stage, or the
         system will not work well on the shop floor.

(3)   Formalise and document maintenance management
         procedures, and make any organizational changes.

(4)   Maintenance management training of supervisors.

(5)   If not already available, draw up
                    equipment lists
                    priority lists
                    p.m. program lists
                    parts lists

(6)   Install software package, test with sample data.
         (If computer not already available, order and
          and install computer prior to this).

(7)   DP User Training of data entry clerks and
         planners.

(8)   Commence implementation.

Presented at the SME Real Time Factory Control Conference, June 1983

Real Time, Plant Floor Machine Monitoring and Maintenance Dispatching System

By Richard W. Murray and W. T. Lesner
Cadillac Motor Car Division

BACKGROUND AND SYSTEM OVERVIEW

The Cadillac Livonia Engine facility is approximately 750,000 square feet in size. 500,000 square feet are part of a building addition which was constructed to house the majority of the machine tool equipment. The plant will produce major engine components (i.e. cylinder blocks, cylinder heads, crankshafts, connecting rods, pistons, manifolds, etc.). It will also house final engine assembly and test.

In the summer of 1978, when plans were being made for the Livonia plant, addition, Cadillac examined many state-of-the-art manufacturing methods. From that analysis, a decision was made to implement a number of advanced applications, one of which is the plant floor information system described in this paper.

Although nearly all of the machining equipment in the plant has been purchased with programmable controller control, it is important to realize that the equipment is geared toward a single purpose high volume production. This means that the information system is different than that found in many other computer aided manufacturing applications...where flexible numerical controlled equipment is ued to produce a variety of different components. This difference is reflected in the system's design and usage. The system design has been a joint effort between Cadillac Process Engineering Plant Engineering, and Manufacturing, along with the Manufacturing Control Systems Group at GM Manufacturing Development.

The Cadillac system consists of a five mini-computer networks which functions as two subsystems. The Machine Monitoring encompasses the monitoring, maintenance and host computers. The remaining two computers are dedicated to the Engine Hot Test System. This system configuration is illustrated in Figure 1.

The primary purposes of the Machine Monitoring and Maintenance Dispatching subsystem are collection of machine production information, machine status interrogation and efficient trades personnel utilization. The Engine Hot Test subsystem, on the other hand, is dedicated to Hot Test process monitoring. Although there will be communication between these two systems, this paper addresses each separately.

The following describes primary functions of each computer in the two subsystems. Detailed information -- in terms of system components, philosophy and back-up features -- is presented in the remainder of the paper.

Monitoring Computer

The monitoring computer's function is data acquisition from the plant floor. The five functions described briefly below are the heart of the plant-floor information system and will be described in greater detail in other sections of the paper.

1. Block line monitoring...
   Monitoring of all the machines and automatic storage units on the
   cylinder block line. The entire line is equipped with 29 Allen-Bradley
   programmable controllers (PCs). Communication with these PCs will be
   done with Allen-Bradley's 'Data Highway'.

2. Head line monitoring...
   The monitoring of 28 Modicon Programmable Controllers on the cylinder
   head line. In this case, Modicon's Modbus will be used to communicate
   within the system.

3. Critical Alarm monitoring...
   Monitoring the status of critical conveyors and filtration units throughout the plant and alerting maintenance through an alarm system when they
   malfunction.

4. Data entry stations...
   Allow operators on the block and head lines to call for maintenance
   assistance or input non-maintenance reasons for downtime from their
   machine control panels.

5. Future expansion...
   Expansion of the machine monitoring concept to encompass the whole plant
   is already underway. The boreliner and conn rod lines will be the first
   additions.

Maintenance Computer

This computer is used to support maintenance activities, as well as the
unrelated function of engine component verification. The primary functions
of the computer are described below, and will be elaborated on in other
sections of this paper.

Cadillac Livonia has decided to implement a centralized dispatching system
for all skilled trades. The centralized dispatcher will track maintenance
activities via a dispatch console which consists of:

1. Critical Alarm display...
   A color CRT which displays the status of critical alarm points and data
   entry station requests for maintenance assistance.

2. Critical Alarm printer...
   This printer goes hand in hand with the Critical Alarm display. It is
   used to obtain hard copy documentation of every alarm or data entry
   station request.

3. Trades display...
   A color CRT which indicates the current assignments of all skilled
   trades personnel.

4. Job backlog display...
   This color CRT is used to keep track of the maintenance jobs being worked on and those that have not yet been completed.

5. Interactive CRT...
   This black and white display, with its associated keyboard, is used by the dispatcher to communicate with the computer. All dispatcher data entries and responses are via this CRT.

6. Communications equipment...
   A standard telephone and special one-way paging system will be the primary means of contacting the tradespersons.

In addition to the maintenance functions previously mentioned, the computer supports a planned maintenance system which focuses on scheduling of routine maintenance jobs to prevent major breakdowns. The system encompasses jobs for all skilled trades including lubrication, electrical, hydraulic, pneumatic and general building maintenance.

This computer also performs Engine Component Verification (E.C.V.). E.C.V. has been used within the auto industry for a number of years to assure that the proper emissions-related components have been assembled to the engine.

The engine inspector uses a light pen to wand the bar code label from all emissions-related components, as well as an engine identifier. The computer correlates these, checks for proper grouping, and prints out a label which acknowledges proper assembly or highlights mis-assembly.

Host Computer

The purpose of this computer is to generate management reports concerning both Maintenance Dispatching and Machine Monitoring. There are 17 black & white and four color CRT's located throughout the plant used for management reporting. A user friendly display generator is used to create color graphic and tabular black and white displays. A high speed printer is attached to the system for printing any displayed report. The displays are designed to provide the user with dynamic management information to assist in decision making on a minute-to-minute basis. The system also archives historical data for the last year to assist in making longer term decisions. Specific functions and uses will be discussed in later sections.

Engine Hot Test Computers

The two computers are used to support a comprehensive automatic Hot Test procedure. They specifically:

1. Track engines throughout the test system.

2. Communicate with test stands via their Allen-Bradley PCs (again over the 'Data Highway').

3. Provide repair information.

4. Provide management reports.

5. Store data on individual engines for immediate access.

Further details on the exact operation of this system will be discussed in the Engine Hot Test section of the paper.

As stated previously, these five computers constitute two functionally separate systems -- machine monitoring/maintenance dispatching and engine hot test. The following sections focus on machine monitoring via programmable controllers, computer aided maintenance dispatching and management reporting capabilities. This paper will conclude with a review of the engine hot test system then close with remarks regarding the entire system.

FIGURE 1

GENERAL SYSTEM
OVERVIEW

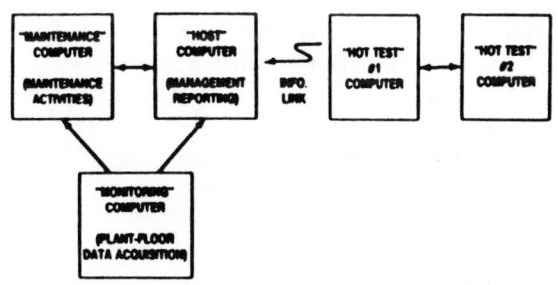

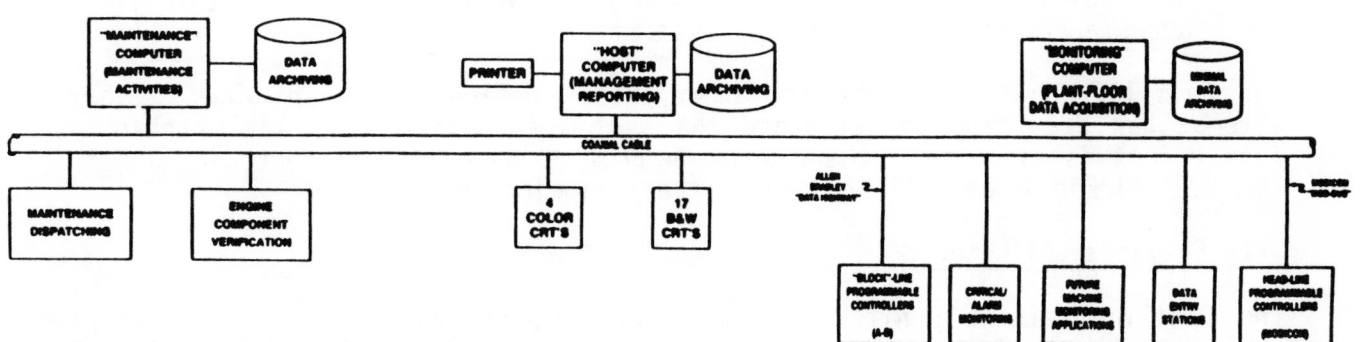

MACHINE MONITORING

Background

Consideration of machine monitoring systems quickly reveals the multitude of approaches that can be taken. In the past, many innovative plants installed monitoring systems that ranged in complexity from hard-wired Telecontrol to mini-computer networks. As a matter of record, a few of these systems provided expected results, some were successful in areas not originally planned for and some were relative failures.

From the initial Cadillac system planning stages, we realized that much could be learned from these earlier systems. What deficiencies caused failure? What features were strong points? What characteristics made them difficult to work with? Study of these systems revealed that many common problems resulted from the approaches taken, some which were simply the result of then current technology. Among these problems were the following:

1. Lack of integrity in collected data. Failures of system hardware often resulted in erroneous data for the associated shift. Unfortunately, these errors often would propogate into day, week, and month totals. This problem was responsible for the death of more systems than probably any other. In most production environments, data inaccuracies may be tolerated once, but repeated errors quickly destroy confidence and interest in a monitoring sytem.

2. Lack of system flexibility. Monitoring system definition involves too many variables and alternative approaches to be absolute. The initial iterations of design typically, are, in fact, learning experiences for the future owner. Is it reasonable to think, then, that these changes and refinements would cease following system installation? Inflexibility to change has perhaps been the second most important factor contributing to unsuccessful systems.

3. Persistent hardware failures. Many past systems suffered from annoying and continual hardware failures. This was due, at least in some part, to the application of immature technology in the industrial environment. Nonetheless, persistent malfunctions are devastating to the credibility, acceptance, and support of any system.

The Cadillac system was designed with these considerations in mind. Every attempt has been made to safeguard the system from these potential problems. This effort is evident in the design -- from PC data collection techniques to computer backup -- as will be seen in the following description.

Data Elements Collected

The Cadillac machine monitoring system is designed to help plant personnel achieve optimal piece production rates from the associated machining lines. As such, the system will serve as a tool in identification of five basic problem areas:

1. At any instant in time, which machines are not producing parts and (as much as possible) why?

2. Are any machines running slow in comparison to their standard cycle time? If a machine is slow what head is controlling the operation?

3. Which machines suffer from -- or are trending toward -- chronic mechanical, electrical, or hydraulic problems?

4. What are the production part counts for this period, and how do they compare to standards?

5. Are there any material flow problems which are causing - or will cause - machines to be idle?

To provide this information, the Cadillac system will collect key data elements from each monitored machine and status points from monitored automation equipment. This collection process will be executed via programmable controllers supplied with machine tools for control, plus some additional units for automation status. In the case of machine tools, each PC will accumulate time and count information in a block of memory registers that will periodically be retrieved by the computer system. The key data elements are as follows:

1. Accumulated machine auto cycle time and occurrence count. Auto cycle time is defined as when the machine is in automatic mode and has no fault conditions to prevent it from cycling.

2. Accumulated machine manual time and occurrence count. Manual time is defined as machine in manual (hand) mode with no fault conditions preventing it from cycling.

3. Accumulated machine fault time and occurrence count. Machine faults are defined as those fault conditions incorporated into the control logic. Examples of such conditions are emergency stop, lube fault, chip drag, probe fault, etc.

4. Accumulated machine cycle time and cycle count. Machine cycle time here represents those cycles which are considered to be normal. Normal cycle time is when the machine cycle is less than 150% of the standard. This separates abnormally long cycles that could potentially corrupt the normal cycle time. Each cycle includes the total piece machining time exclusive of loader or unloader waiting.

5. Accumulated machine excess cycle time and cycle count. Excess cycle time is defined as those cycles in excess of 150% of standard (non normal cycles). Excess cycle time is an indication of long cycles or mechanical problems with the machine.

6. Accumulated machine 'waiting for cycle initiate' time and occurrence count. This waiting time is defined as those intervals in which a machine is capable of cycling (in automatic mode) but requires depression of the 'cycle start' palm buttons. This condition can occur in two situations.

a. the machine is in single cycle mode, or

b. the machine has been switched from manual or single cycle to automatic.

7. Accumulated machine waiting for parts time and occurrence count. This waiting time is defined as those intervals in which a machine is in automatic and clear to cycle, but cannot due to lack of material.

8. Accumulated machine waiting to unload parts time and occurrence count. This waiting time is defined as those intervals in which a machine is in automatic and clear to cycle but cannot due to material flow blockage (back-up) at the unloader.

9. Accumulated machine idle time and occurrence count. Idle time is defined as the total time in which a machine experiences one or more of the above three states -- waiting for cycle initiate, lack of material, or material unload blockage.

10. Machine part (piece) count. The total number of pieces through the end of a machine.

11. Machine rejected part or probe fault count. These counts reflect the number of out of specification parts detected/rejected at online gaging stations and the number of probe faults which check for broken tools.

12. Unit last cycle time. These time values are maintained for each active station (i.e. head) and total cycle for each machine.

13. Tool change counters. The Cadillac plant uses two counters per machine for tool change frequency and replacement program duration. These counters are maintained by the PCs and, it may be noted, are viewable from a display panel at the machine control console.

14. Gage classification counts. Classifier counts are used to record the number of parts within several predetermined dimensional windows. This information is used for statistical quality control in monitoring critical dimensions.

15. Torque classification counts. These counts correspond to results registered from monitored fastening (i.e. nut-runner) stations.

The above data elements are accumulated, as previously stated, into PC memory registers for subsequent retrieval by the computer system. In addition to these timers and counters, status information will be assembled into memory word 'bits' to provide current machine status summaries.

Data Collection Tecniques

Using programmable controllers for machine control provides an excellent data collection vehicle for the monitoring system, a vehicle not available with relay controls. The PC, by definition programmable, can generally be 'fitted' with data collection logic to monitor any input, output, or logic coil desired.

When designing the Cadillac system, however, we realized that certain guidelines (techniques) should be followed to ensure high data integrity and system manageability. These guidelines were as follows:

1. PC data collection logic should provide stand-alone accumulation for 24 hours.

2. Data collection logic should be consistent from one PC to another.

3. Data registers should form a continuous block within each PC.

4. PC logic line counts (space required) should be held to a minimum.

The first guideline is perhaps the most important, for it establishes data buffering at the PC level. As stated previously, a basic goal in the Cadillac design was to ensure data accuracy at all times. To ensure this, it is desirable to provide data buffering at lower levels of a system hierarchy, such that hardware failures at higher levels will not result in data loss. For this purpose the data collection logic in each PC was designed to ensure 24-hour stand-alone accumulation. The primary implication of this capability was that time and count registers provide sufficient 'width' or capacity. In the case of timers, a 0.1 second time base was chosen to coincide with machine supplier cycle timing conventions. A 24-hour accumulation of time, such as 'autocycle' would then require storage of 864,000 (24 hrs. x 36,000 tenths) 1/10 second second increments. For the Allen-Bradley and Modicon PCs used on the Cadillac system this storage required double precision (cascaded) registers. Counters, on the other hand, demonstrated less severe capacity requirements. As a guideline we considered that the fastest cycling machine would produce one part in 15 seconds, thus requiring a range of 5760 (24 hrs. x 240 parts per hour). For Modicon PCs this implied single precision counters, for Allen-Bradley double precision.

Figure 2 (on the following page) illustrates manipulation of auto cycle timer and counter registers by an example PC ladder segment. Note that in this case a double precision timer is used to provide the desired range.

Consistency of logic from one PC to another was a goal to simplify system implementation and future logic maintenance. Several practices were followed to ensure this consistency. From the start of system design it was planned that timing/counting ladder logic would be conceptually identical within PCs of the same type. This commonality will generally promote troubleshooting and logic maintenance ease. Another practice was that monitoring logic would be entered into each PC at the 'end' of user memory area. This practice ensured a physical separation of control and monitoring logic lines to avoid confusion arising from interleaved logic.

The third monitoring logic guideline -- contiguous data registers to form a block -- was pursued to provide simplicity and efficiency in register acquisition by the computer system. These benefits result from a peculiarity of most PC-computer interfaces -- register acquisition requests are serviced once per scan and involve sequential memory locations. In order to minimize the number of acquisition transactions -- and thereby the software and transmission overhead -- desired registers should be 'stacked' as contiguous blocks. Multiple requests generally would require multiple PC scans, with the time delays of 25-75 milliseconds per scan. One can readily see the potential for undue data acquisition delays.

The last monitoring logic guideline was that PC space requirements be held to a minimum. This goal was to function of reality as much as good design practice At an early design stage it was recognized that some PCs would offer limited free space, and still provide reserve for future logic modifications, the monitoring logic was optimized to reduce line count.

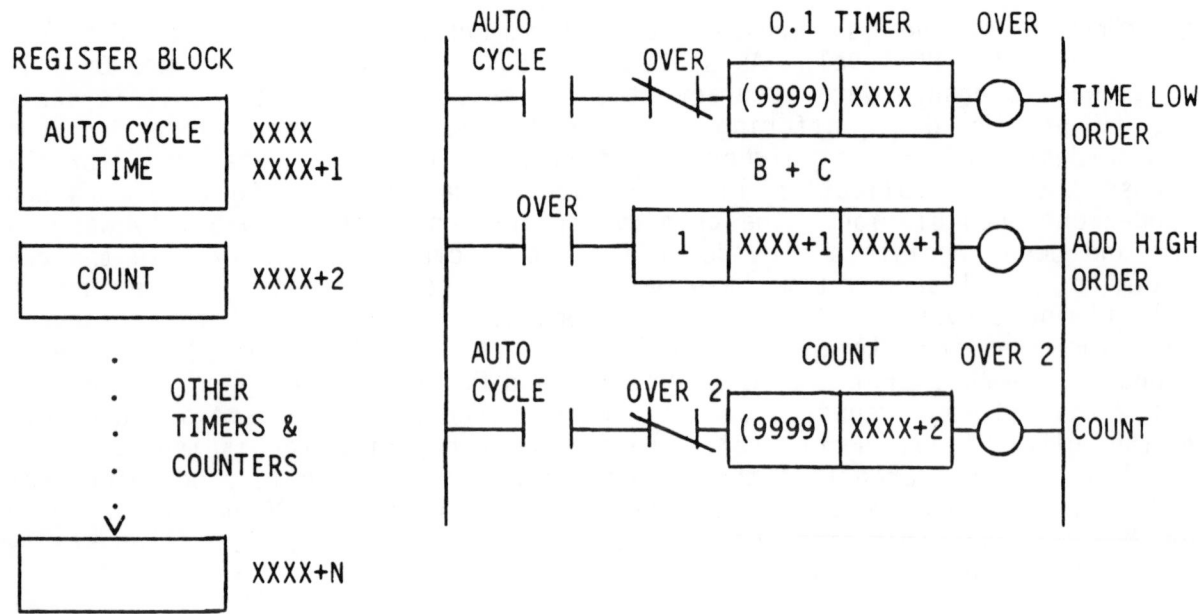

Figure 2

ACCUMULATING DATA ELEMENTS

The above guidelines resulted in PC data accumulation techniques that ensure a
in PC data accumulation techniques that ensure a high integrity, efficient, and
maintainable collection process. It should be noted that Cadillac was in the
favorable position of defining to machine vendors what logic would be supplied
with the machine tools. The timing of machine delivery was such that this
could be done, as opposed to adding monitoring logic 'in the field'. This
ability can greatly simplify system implementation and should be pursued whenever possible.

PC Computer Networks

Programmable controller computer interfaces have made possible a variety of
systems, ranging from ladder logic to process control networks. Just a few
years ago the interfaces available were discrete line (one-to-one) types where
each PC was directly connected to a computer on a dedicated link. While this
discrete line interfacing met the needs of many systems, it was inherently
cumbersome and costly where many PCs were involved.

Production monitoring systems generally are of this type -- the Cadillac
systems initially will encompass 65 PCs with expansion planned to double that
count. The quantity of wiring required by discrete computer - PC links and
the number of computer serial channels required would be prohibitive.

The recent availability of multi-drop (shared) PC to computer networking hardware has provided an efficient communications alternative. It allows connection of multiple PCs (up to 64 in some cases) to a single serial link. This
approach, as illustrated in Figure 3 places serial port and cable length
requirements into a manageable range. In addition, future expansion is generally easier by only requiring extension of the nearest link.

Figure 3

PC INTERFACING SCHEMES

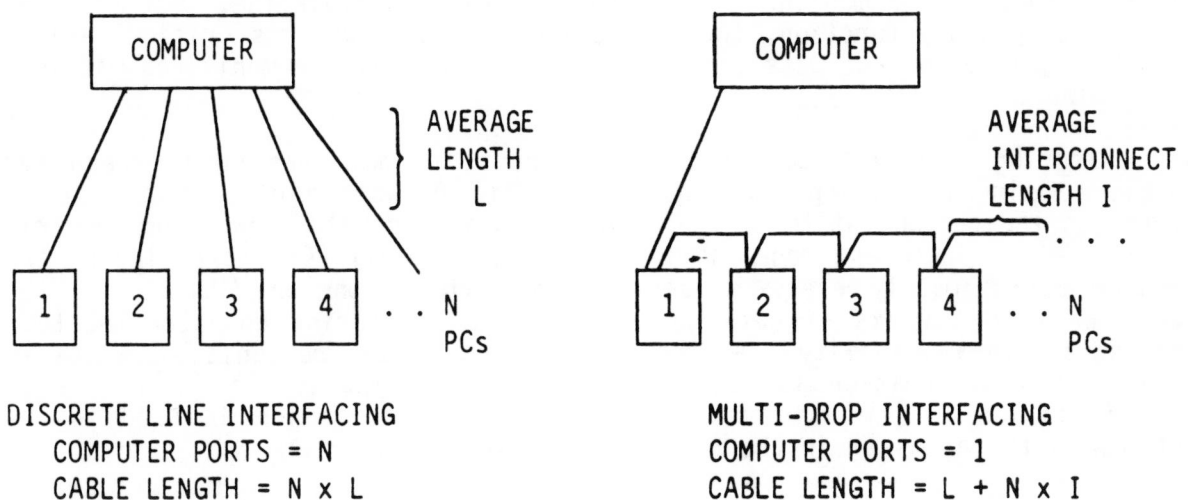

DISCRETE LINE INTERFACING
  COMPUTER PORTS = N
  CABLE LENGTH = N x L

MULTI-DROP INTERFACING
  COMPUTER PORTS = 1
  CABLE LENGTH = L + N x I

The Cadillac system will use Modicon's Modbus and Allen-Bradley's Data Highway for these PC-computer networks. Both products involve multiple PC interfaces interconnected on a common cable with a computer interface for data acquisition. A typical configuration of these networks is shown in Figure 4. The manufacturers support their associated communication line with application software to simplify the monitoring process. The software provides basic utility routines to handle I/O transactions with error detection and recovery. Each Modbus network operates at 19.2 KBPS (1000 bit per second) and supports up to 64 PCs. The Cadillac system will use one Modbus and three Data Highway networks.

Figure 4

PC-COMPUTER INTERFACE HARDWARE

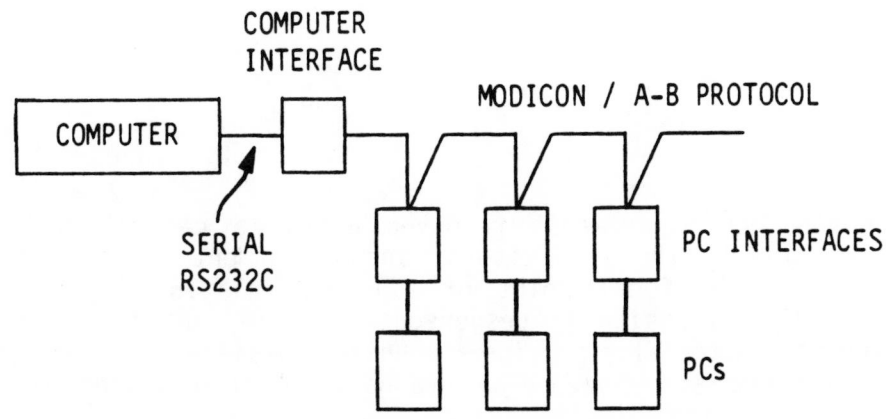

Monitoring Computer

The monitoring computer is the data collector for the computer system. Its primary purpose is to monitor the PCs on the head and blockline. But in addition to its primary function, the monitoring computer supports critical alarms throughout the plant and data entry stations located on the monitored machines in the plant.

The monitoring computer accumulates data about the machines on the head and blockline using a polling technique. The status word mentioned previously determines the current state of the machine and an old value/new value comparison is used to increment mode timers and counters. All PCs are polled by the computer approximately every 30 seconds to maximize the credibility of the data. We keep 2 weeks of data buffered at the monitoring computer level to increase system reliability. As the monitoring computer recognizes changes in the PCs this information is passed onto the host computer which then stores the information in the main database. How this information is used in the system will be discussed in the host computer section.

The second function of this computer is monitoring critical alarms throughout the plant which includes overhead conveyors, filtration units and chip conveyors. Digital communication devices are used to monitor these points and is serially linked to the monitoring computer via RS-232 ports. The communication network uses a daisy chain network to tie the critical alarm points together as well as to the computer. The computer monitors the status of these points and as is it recognizes changes in status passes the changes onto the maintenance computer. The application of these status points as part of the system will be discussed in the maintenance computer section.

The last function of the monitoring computer is interacting with data entry stations (DES) on the head and blockline. DES have a dual purpose of allowing the machine operator to input reasons for downtime or request maintenance assistance from his machine. Typically, each monitored machine has a DES attached to its control panel. When the machine is turned off, the monitoring computer automatically accumulates downtime during this period. The DES is used by the operator to enter standard reasons for this downtime. Five consecutive cycles clears the downtime reasons and stops their respective timers. The operator can also use the DES to call for maintenance assistance. Via a prompting sequence, the operator enters a standard code for the trade required then enters a standard job description. Lights on the DES provide the operator with feedback to acknowledge that the job was opened and finally when a tradesperson was assigned.

MAINTENANCE DISPATCHING

Background

Traditional maintenance departments revolved around a central logbook, hand written notes carried by the supervisors, and management input. The maintenance supervisor is the focal point of all job related activity. He is responsible not only to assign tradespersons to jobs, but track progress of the jobs, expedite material, provide technical assistance, and coordinate multiple skilled trade activity. The drawback of this system is that the supervisor should be in several places at once to do an effective job. He had to be in the shop to keep track of the maintenance jobs and assign tradesmen. At the same time, he should be out on the floor as a technical advisor and to coordinate parts as needed. Over time, many maintenance organizations recognized the inefficiencies associated with the system described previously. This is why many maintenance departments evolved into manual maintenance dispatching systems.

Manual maintenance dispatching systems recognize the need for a single person located in the shop to keep track of all jobs, tradesmen, and the priorities throughout the plant. Based upon the latest information, he can make systematic decisions as to where the tradesmen will be assigned, what jobs will be covered and how fast. This allows the supervisor to focus his attention on actually repairing the equipment, rather than managing human resources. The dispatcher tracks jobs and manpower in the system by using open job and manpower cards. Each time a job is called in, the dispatcher fills out a job card, then when a tradesperson is assigned the job and manpower cards are mated for the duration of the job. This system is effective but burdensome for the dispatcher with regard to opening jobs and verifying equipment related information during peak periods of the day.

With the advent of manufacturing mini-computers, one of the first applications was to computerize manual maintenance dispatching, burdening the computer with keeping track of the status of the jobs and tradesmen. The computer replaced the cards that were used to track the jobs and manpower along with the books that previously contained the equipment related information. Inherent in many computer-aided dispatch systems, historical information is automatically stored about work performed by tradesmen on machinery in the plant. The early systems were very basic, often parallelling the capabilities of the manual dispatching systems. As they evolved over time to meet the particular needs of the organization, several criterion for a successful system became clear. These attributes are listed below.

1. User-friendly and facilitate dispatching...
   It must be easier to use the computer than dispatch manually.

2. Instantaneous response time...
   Users of the system will lose faith if the system is slow.

3. Zero computer downtime...
   Computer downtime has been the cause for the fall of many systems. If manual backup systems have to be used often, the computer becomes more of a burden than a help.

4. Communication...
   There must be stable lines of communication between the dispatcher, tradesmen and maintenance supervisors.

5. Management support...
   Maintenance management must believe in the concept and support computer-aided dispatching.

These five qualities were used as guidelines in developing the Cadillac computer aided Maintenance Dispatching system.

Cadillac's Dispatching Methodology

The Cadillac computer-aided Maintenance Dispatching system was designed to utilize state-of-the-art technology, and the experience of other dispatching systems, keeping in mind the needs of our own maintenance organization at the same time.

Our concept of Maintenance Dispatching revolves around a central dispatcher located in the maintenance shop. The dispatcher uses three color CRTs, an interactive black and white CRT, and an alarm printer. The first color CRT is used to show the status of critical alarm conveyors and inform the dispatcher when a data entry station request is made for maintenance assistance. When either of these conditions require the dispatchers attention, the square on the screen begins to blink, and sounds an alarm. The alarm printer goes hand-in-hand with this color display. It is used to obtain hard copy documentation of every alarm and data entry request. The second color display indicates all available skilled trades and their current assignments. Color is used to high-light key information regarding each trade's status. The third color display is similar to the trade's display, except it is used to show maintenance jobs that are not yet complete...this is the job backlog. This display informs the dispatcher of the status of any job in the system. The screen is arranged by priority, where color has been used to assist the dispatcher in quickly identi-fying key information about each job. A black and white interactive display is used by the dispatcher to "talk" to the computer.

Cadillac's first goal in designing the Maintenance Dispatch system is to make it be as user friendly as possible and facilitate dispatching. For this pur-pose, the interactive portion of the Maintenance Dispatch system is primarily menu driven to simplify the operation of the dispatch station. The dispatcher uses a central menu which displays all functions available to the dispatcher and allows him to move to the functions listed below.

1. Open job...
   Using machine number as the identifier, the dispatcher then verifies the computer supplied information and fills in the caller's name, trade required, and job description.

2. Assign man...
   This function allows the dispatcher to assign tradesmen to jobs in the system.

3. Modify job...

   Is used to alter an existing job in the system.

4. Close man...
   After the tradesperson finishes with the job, he tells the dispatcher the status and is then assigned to the next job.

5. Close job...
   When the job is completed, the dispatcher enters a description of what repairs were made to the machine and then closes the job and the associated man.

6. Edit equipment...
   Allows the dispatcher to alter the machine related data.

7. Edit manpower...
   Allows the dispatcher to alter the information associated with each tradesperson.

In addition to using the central menu to move from function to function, the dispatcher can use programmable function keys to expedite his job.

The second goal in designing the system is to provide the dispatcher with instaneous response times. Cadillac has dedicated a mini computer to the primary purpose of supporting Maintenance Dispatch. With Maintenance dispatching having such a high priority, response times to the dispatcher are very quick

The third goal of the Cadillac system is to eliminate Maintenance Dispatching downtime. Downtime is eliminated by designing the system so the Maintenance Dispatching can be backed up on the computer which normally generates reports. In other words, both maintenance and reporting software will run on the host computer. These systems have been designed so that performance of either system will not be affected when running in backup mode.

Cadillac's fourth system goal recognizes the need for lines of communication between the dispatcher, tradesmen, and the maintenance supervisors. The dispatcher is the focal point of all communication in the system. Floor personnel interface with the dispatcher using telephone located throughout the plant. All tradesmen wear one-way pagers which allow the dispatcher to contact tradesmen at any time and give them a brief message or have them call the maintenance office. Maintenance supervisors wear two-way radios with the dispatcher controlling the base station. The system is designed so that transfer of information is easy for everyone involved. The dispatcher relies on this information, to aid him in making the day to day decisions with which he is faced.

Figure 5

LINES OF COMMUNICATION

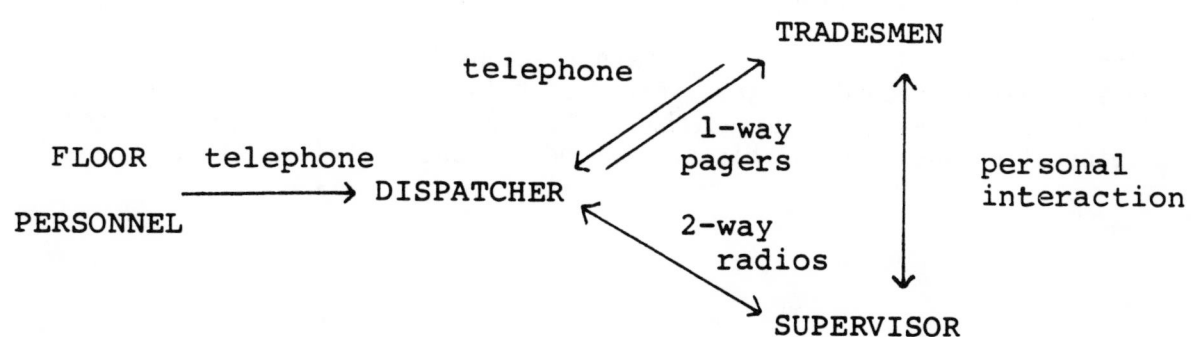

The last goal and most important factor contributing to the success of our Maintenance Dispatching system is that our maintenance management has been involved in the design and implementation of the system from the very beginning.

Their input has been very helpful to design a successful system that meets their needs and the needs of the maintenance department.

Whistles and Bells

The Cadillac Maintenance Dispatching System surpasses all others because it is a total system. It attempts to handle those tough situations that have been ignored by many other maintenance systems. The Cadillac system addresses challenges like historical information, accounting support data, non dispatched work and shift change.

Historical information is a fringe benefit of computer-aided dispatching systems. We saw this as "the sleeping giant" that could potentially provide us with the tools to make great strides in machine uptime. For this reason, we keep the last year of job and trade related data on-line, available to be displayed at any time via the host computer. Information like all the jobs performed on a machine over a period of time could be invaluable in determining repetitive breakdowns.

Accounting information was built into the original design of the Maintenance Dispatching system. The computer keeps track of all jobs each tradesperson works on and the time accrued to each job. This information is then used to print non-productive time tickets that are used by financial to charge maintenance time to productive departments.

Non-dispatched work becomes more important when you consider historical data as a tool to predict breakdowns and that tradesmen are literally being paid by the time tickets the computer is generating. Non-dispatched work occurs in many situations such as those listed below:

1. Weekend or holiday work...
   A situation where there are only a small number of tradesmen working.

2. Resident work...
   Some areas require the constant attention of a skilled tradesperson.

3. Work done without the dispatchers knowledge...
   Minor jobs worked on in the plant outside the dispatching system.

Cadillac has chosen to deal with these situations by having the tradesmen and/ or the supervisors record the jobs worked on outside of the dispatching system. On the shift following non-dispatched activity, the maintenance clerk enters this information updating all historical and accounting information.

212

Shift change is perhaps the busiest time of the day for the dispatcher and perhaps the most important in terms of data collection. In a period of a half hour, one dispatcher is closing all his people and another is assigning his, in addition, all the jobs are being lined up for the following shift. We allow the dispatcher to pre-close trades up to a half-hour before the end of the shift, then at shift change, all pre-closed trades will be removed from their respective jobs by the computer. To facilitate the job of the incoming dispatcher, we allow him to preassign his trades up to a half-hour before the start of the shift. Even though the dispatcher has assigned the trades the dispatcher has assigned the trades the computer will wait until the start of the shift to actually assign them to the jobs.

Planned Maintenance

Planned Maintenance (P.M.) is a concept that everyone believes in, but in reality, it is often pushed aside in order to service major breakdowns. A lack of up-to-date information complicates the process because the P.M. information is stored away in manuals and/or no one is sure when the machine was last serviced. In many cases, a machine related failure will be the event that signals the need for this type of maintenance. Cadillac's plan to meet this challenge is to design a method for scheduling routine jobs over three shifts, print the schedule each week and then distribute a list of routine jobs to the P.M. tradesperson daily. The system is made up of a P.M. database, scheduling program, and operators interface which are all a function of the Maintenance Computer. The P.M. database is made up of three types of information which are machine, P.M., and scheduling.

The scheduling program interacts with the database then using the schedule, prints all routine tasks for each of the next seven days and all infrequent tasks on a weekly schedule. This is printed once a week and then distributed on a daily basis to the skilled trades. The operator interface is used to make changes to the P.M. tasks and the scheduling data. These changes are made based upon feedback from the tradesmen servicing the jobs. All changes would then be reflected in the next printing of the schedule. A tremendous amount of flexibility was built into the system so that as the needs of the plant change, the P.M. scheduling system could bend to meet those needs.

HOST COMPUTER...MANAGEMENT REPORTING

Background

Computerized management reporting has been around for a long time. Data processing generates bundles of reports for distribution every day. There is an abundance of time sharing systems that perform statistical analysis, tedious record keeping, and print charts and graphs. So what is different about the way Cadillac displays the information associated with the Machine Monitoring Maintenance Dispatching computer system.

-   Dynamic information...
    The information you are looking at represents what is happening at this moment.

- Middle Management...
  Now has the tools to manage their human and mechanized resources on a minute-to-minute basis.

- Display graphics...
  Each production supervisor uses a color graphic terminal to view the status of his entire machining line at a single glance.

As you can see Cadillac's interpretation of management reporting is very much oriented toward providing operating levels of management with dynamic information for decision making. Since the majority of the system revolves around real time reporting, it is extremely important to minimize downtime of the reporting system. We have built into the system configuration the ability to run the reporting system in back-up mode when the host computer is down. Back-up mode for the entire system and an overview of the system configuration will be detailed later in this section. Flexibility and ease of programming graphic and tabular reports is another "plus" of the system. An english based report generated is used to design displays/reports for Machine Monitoring and Maintenance Dispatching information stored in the database.

System Configuration

A significant characteristic of the Cadillac System is the use of broadband coaxial communications. At an early design stage, it was decided to utilize broadband cable that was previously installed for a plant fire and watch system Broadband is high speed, multiple user and supports a wide range of information transmission from digital to video.

In the Cadillac system, the PC networks, computers, plant floor terminals, and data entry stations are connected to the broadband cable. This interconnection is illustrated below in a simplified manner (Figure 6).

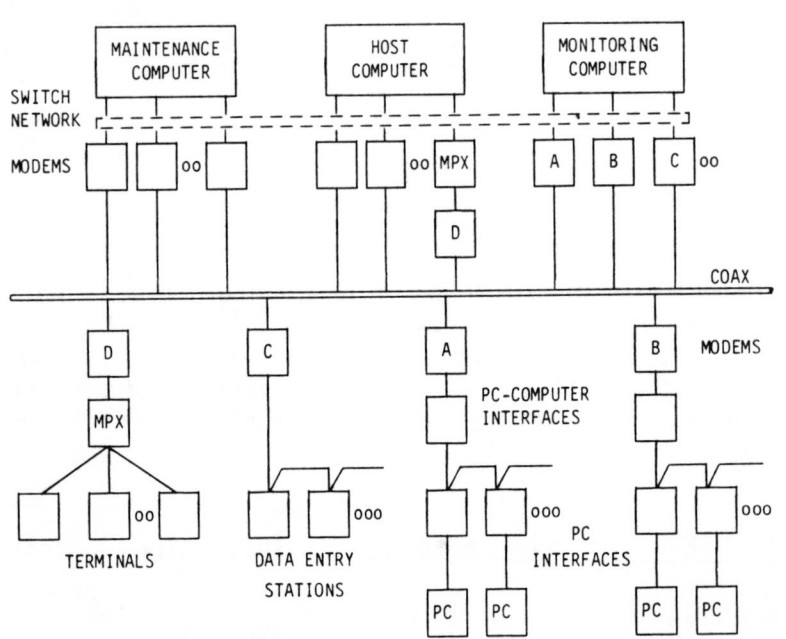

Figure 6

SYSTEM COMMUNICATIONS

The boxes immediately to either side of the cable are broadband modems, some of which illustrate frequency (channel) selection by letters A, B, C. Also shown are the PC-computer network interfaces connected to the broadband via modems. The boxes labeled MPX are statistical multiplexers used for the plant floor communications. These provide multiplexing of multiple terminals (up to 8) over a single serial line. The Cadillac system will initially use 21 floor terminals with a significant advantage obtained through multiplexing.

Use of the broadband system will provide Cadillac with considerable flexibility and interconnection ease, particularly in future expansions. This technology, plus multi-drop PC networking and terminal multiplexing, provide system efficiencies not possible a few years ago.

An efficient method of computer backup is to interconnect the primary and secondary processors with high speed data links. These links are used for data base transaction notification and fast data file transfer for smooth control transactions. Although the broadband network is high speed on the trunk cable, the speed of most modems is restricted to 9600 BPS. This speed is not sufficient to effectively handle mass data transfers. Because of this restriction, the Cadillac system will utilize a separate network between computers as shown in Figure 7.

Figure 7

COMPUTER-COMPUTER NETWORK

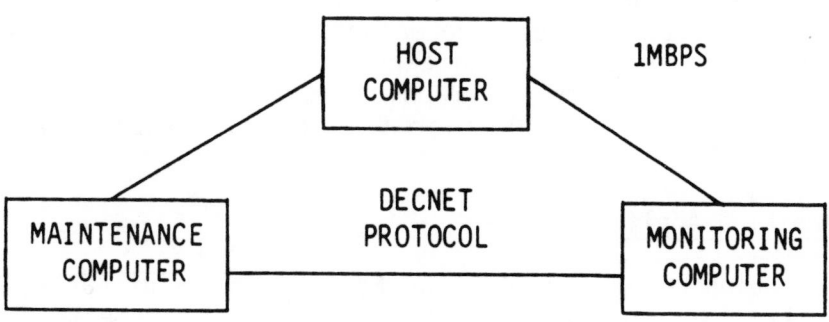

Due to the fact that Digital Equipment Corporation (DEC) computers are in the system, it was decided to use a DECNET DMC11 interface network. This network is capable of running at 1MBPS and is supported by DECNET protocol software.

The DECNET communications system will be used for time critical transmissions such status messages and data file transfes for backup. In the Cadillac system each computer can backup another computer as per the following:

1. The maintenance computer can backup the host or the monitoring computer.

2. The host computer can backup the maintenance or the monitoring computer.

3. The monitoring computer can provide limited report capability in the event of maintenance and host computer failures.

It should be noted that plant floor devices are 'shared' between the computers for backup purposes. This is done by switching the subject modems from the primary processor to the secondary when a backup transaction occurs. The reverse is executed when the primary once again gains control. The Cadillac system uses multi-channel switcher networks for this modem transfer.

Report Generator

One of the most mysterious and frustrating qualities about computer aided reporting is the mystique associated with the displays and reports themselves. Very few users of the system really understand how a report is written, let alone actually write one using real time basic or fortran. Typically, the original reporting structure of most systems meet the needs of the users but as time goes on those needs change. Many times the system is not flexible to change, or there is no one with the technical expertise to make the changes. This point was important enough in Cadillac's opinion, to design into the system a non-technical user oriented graphic display and report generator. Cadillac's display generator uses english commands to design the displays and predefined descriptive names to retrieve database items. This allows the logical user with little or no programming experience to modify existing displays and/or design his own. The color graphic protion of the package uses the graphic capability OF THE ISC 8000 color CRT to draw graphs, bar charts and even pie charts.

Management Information Applications

Management information from the Machine Monitoring Maintenance Dispatching system is typically used by two separate groups of management. The Machine Monitoring information is primarily used by production hourly employees, supervisors, and assistant superintendents to aid them in managing their human and mechanical resources to most effectively produce a cost competitive quality product. On the other hand, Maintenance Dispatching aids maintenance management in allocating manpower to breakdowns in a systematic manner and at the same time provide them with the historical information to predict breakdowns in the future. Even though the information on the system seems to be geared toward these two specific groups, all of the information within the system is available to whomever may be interested.

In the original concept of Machine Monitoring, Cadillac's goal was to give the people out on the floor the tools to better manage their machining lines. Listed below are some of the displays available that refer to the monitoring of the head and blockline.

1. Color graphic line status...
   Is used by the production supervisor to view the entire status of the machining line. All the machines on the line are graphicly represented, color is used to highlight the status of each machine. Under each machine is performance related information for this shift.

2. Color graphic operation status...
   Is similar to the linestatus display, but focuses on an individual machine of the line. Each station and active head of the machine is drawn on the display. Inside each head is a description of the function performed and the last cycle time of that head. Below the graphic display of the machine is performance related information including piece counts, efficiencies, tool change, and maintenance information.

3. Color graphic history...
   Designed to use the graphic capabilities of the system to informatively represent machine performance.

4. Float status...
   Displays a snapshot of where all in-process float is in the system.

5. Production history...
   Displays general performance related information to provide the user with an overall view of the performance of the line.

6. Mode history...
   Provides the user with detailed information for each of the four major machine states, running, idle, down, and fault.

7. Idle history...
   Displays downtime information received from the data entry stations on the monitored lines.

Maintenance Dispatching reporting has followed the same line of thought as Machine Monitoring, but obviously the reporting structure is geared more toward the maintenance organization. The following displays are representative of those used by maintenance management.

1. Top 20 report...
   Identifies the top 20 pieces of equipment with the longest job duration, most maintenance calls and most maintenance hours. This report highlights equipment that is extensively requiring maintenance assistance.

2. Completed job history...
   Allows the user to display completed jobs for a period of time sorted by any or all of the variables in the system.

3. Response time summary...
   Looks at the number of dispatches and average response time for each of
   the priorities within the system. This display provides the dispatcher
   with feedback concerning response times to the various priorities of
   equipment in the plant.

4. Calls per person...
   Details the number of maintenance calls for a tradesmen over a period of
   time.

The reports discussed in the previous two sections are baseline reports
designed to meet the initial needs of the users of the system. The user's
imagination is the limit to the combination of information and the ways to
display this information using the color graphic capabilities of the report
generator.

ENGINE HOT TEST MONITORING

Hot Test Overview

The Hot Test line is a palletized, automatic line. Completely assembled engines
are loaded onto pallets. The engine is identified to the computer by bar code
reading the engine's unit number and type. These two engine parameters are
correlated to the pallet number, which becomes the key means of identification
throughout the Hot Test system. The pallet number is represented by a series
of magnets mounted on the side of the pallet, and it is read by a magnetic
reader. These numbers are routed to the PC, which controls the conveyor system,
and from there to the Hot Test computers.

Ultimately, the engine enters one of 14 Hot Test stands, each of which is
controlled by an Allen-Bradley PC. The PC, however, cannot begin the test
sequence until it knows what type of engine to test. This process of engine
identification involves another magnetic reader which recognizes the pallet
number and, via the PC, retrieves from the computer the corresponding engine
type. Given the type, the programmable controller can set the proper test
parameters and begin the test.

When the test is completed, the PC sends the test data to the computer for
storage and future reference. If an engine is sent to repair, the repair-person
can display the engine record and reasons for the engine's failure. Once
finished, that person can enter repair information and where the engine is
being sent -- either back to be retested, out of the system for further repair,
or out of the system as an accepted engine.

When the engine has been accepted, either by the test stand or a repairperson,
it is routed to the unload station where another reader (attached to the
conveyor PC) identifies the engine and verifies that it has passed all the
tests.

Data Collection Process

Data collected from the Engine Hot Test portion of the system is significantly different than that collected in machine monitoring. In machine monitoring, the purpose was to act like an 'outside efficiency expert', analyzing all portions of the machining cycle, but not influencing the operation of the machine. In Engine Hot Test, the computer is an integral part of the system, and the goal is not to analyze efficiency, but to ensure quality.

Each data collection source (the conveyor programmable controller, Hot Test Stand PC, and repair terminals) will be discussed briefly. The specific interaction between the PC and the computer will be covered under 'Data Collection Techniques'.

1. The conveyor PC...
   As was mentioned in the overview, the conveyor PC coordinates the initial engine identification process at the conveyor load station and the final verification at the unload station. The bar code reader is interfaced to the PC by means of a serial computer link. The magnetic readers are backed up by manual keyboard entry).

   When an engine is loaded onto -- or unloaded from -- the conveyor, the PC collects the proper information from the bar code reader and/or magnetic reader and passes the information to the computer. In the case of the unload station, the computer returns the status of the engine (accepted, additional repair necessary, retest, or unknown) which the PC uses to direct the load operator as to the disposition of the engine.

2. The test stand...
   The programmable controller on the test stand communicates with a variety of devices. In addition to communicating with the pallet's magnetic reader and the operator's console, the PC must also collect large amounts of test data directly from the test stand (through analog and digital I/O), as well as via a microprocessor (through a serial computer interface) which accumulates high speed test information.

   For eight different RPM/Load combinations, the PC dynamically collects test data on items such as oil pressure, torque, exhaust pressure, vacuum etc. and compares it to preset high and low limits. Also, certain items of data such as power contribution are broken down by cylinder.

   In addition to the dynamic analog data retrieval, the PC also performs pass/fail checks on a number of items such as water cavity leaks.

   At the end of the test cycle the PC, which has stored these values in its registers, transmits the test information to the computer. The PC must get a verification that the computer has received the test data before it can begin the next test.

Other than automatic test data, the PC must also:

   a. Perform the engine recognition procedure so that it knows what limits to set during the test cycle.

b. Keep track of the cycle time of the test.

c. Accept any operator inputs (such as a visual reject) which may occur during the test cycle.

3. The repair area...
   As was mentioned in the overview, the repair person can see the test history of the engine via a repair CRT mounted on the repair stand. Data is sent to the computer from the repair CRTs using current loop transmission at a data rate of 2400 BPS.

   Only two types of data are collected from the repair area. The first is repair information. The operator can enter five different repair codes (selected from over 100) and a comment as to what was done to repair the engine. This is kept as part of the permanent record for the engine. The second item of data is the engine disposition (where the engine is being sent). This item is necessary to be able to track where engines are in the system.

Data Collection Techniques

In this section, the techniques for communicating between the PCs and the computer will be highlighted. There are two main distinctions between the PC/computer interaction of the Hot Test system and that of the Machine Monitoring system.

The first difference is the polling technique used to obtain data from the PCs. In the case of Machine Monitoring, the computer solicits all requests for data and thus has control over what (and when) data is obtained. Engine Hot Test, on the other hand, utilizes a floating master concept, in which the PC sends its data to the computer without any prior solicitation.

When a PC has completed the test cycle, the data (128 words for both accepted and rejected engines) is sent to the computer. The computer task, which receives the data, runs at a high priority. Although data can be received from any PC, only one PC is serviced at a time. This means that all of the data is read by the computer to a buffer for that particular PC, it is verified, and then the computer responds to the PC with the appropriate protocol. During this data transfer process, the PC is allowed to release the engine, but cannot start a new test cycle until the data has been received and acknowledged by the computer(s).

Other test stands that have completed their test cycle are queued and serviced in the same manner, on a first-in/first-out basis, until the previous acquisition is complete.

The second difference is the way that the computers are backed-up. The Machine Monitoring concept discussed previously utilizes a switching network for determining which computer polls the machines. The data obtained by the monitoring computer is then shared with the other computers over a high speed (1 MBPS) communications link.

In the Engine Hot Test system, the repair terminals are handled the same way, utilizing a switching network. The PCs, on the other hand, send messages to both computers over the 'Data Highway'. In fact, in order for the PC to begin the next test, it must have received an acknowledgement signal for both computers (when both are operational). This assures that when both computers are running, they are obtaining identical information. When one computer fails, the other computer records all the information during that period of time. When the failed computer is functional again, the unrecorded information is transferred over the high speed link until its database is up-to-date.

Hot Test Reporting

Just as in the Machine Monitoring/Maintenance Dispatching portion of the system the Hot Test management reporting capability is considerable. Production status and historical reports, reject and repair summaries, as well as equipment and repair efficiencies, are all available on demand.

In addition, the last 20,000 individual engine records are on-line for immediate access. Information about any engine, while it is being repaired or going through final car assembly, can be obtained with a minimum of delay.

CONCLUSION

The plant-floor information system outlined in this paper provides a good starting point for the development of a total manufacturing information system at Cadillac. Obviously, as one considers the vast additional applications which can be integrated into the system (such as energy control, material control, and time and attendance) it becomes evident that any information system must be built to evolve with the needs and levels of sophistication within the organization.

This need for flexibility has been recognized and designed into the system wherever possible. This is evident in the use of multi-user broadband coaxial cable, the high speed inter-computer data links using DECNET, and extra capacity at each mini-computer.

Furthermore, a computer system which interfaces with plant-floor personnel should not be implemented in an isolated manner; it must be well planned and integrated into the overall operation of the plant. The best PC/computer interaction and information gathering techniques are useless unless the people who use the information are taken into consideration. This is especially true since computerized information systems have a tendency to generate feelings of mistrust and de-humanization. With this in mind, Cadillac has maintained plant floor supervision involvement in system design from initial stages. In addition, both management and union representatives have been informed of the design detail and project progress.

Finally, it is hoped that this paper has served, not only to highlight the technical data gathering features of the system, but that it has also accentuated the importance of transforming raw data into valuable management information.

Presented at the Refinery and Petrochemical Plant Maintenance
Conference in San Francisco, February 1979

Development & Application for Computerized Maintenance Scheduling & Control

By A. David Leone
Sun Petroleum Products Company

During the presentation, I will address the development and application for computerized maintenance scheduling and control. Further, the presentation will address the development and application of a computerized preventive maintenance program. Results achieved thru the use of the aforementioned computerized programs will be discussed.

Specifics of the computerized maintenance scheduling and control system will include: equipment utilized, preparation of normal maintenance schedules, development of management reports, use of descriptors, normal and capital backlog control, equipment distribution, development of garage and shop schedules and results achieved.

The preventive maintenance program discussion will address the approach taken for field input, selection of equipment covered, methodology for determining the amount of PM, PM techniques applied, method of issuance, and results achieved.

COMPUTERIZED MAINTENANCE SCHEDULING AND CONTROL

Approximately two years ago, Sun Company's Marcus Hook Refinery began to seriously consider the employment of mini-computers for day-to-day planning and scheduling activities. Obviously, in the past, like other refiners, Sun's Marcus Hook Refinery had utilized computers for large projects(shutdowns, large jobs, etc.) utilizing computerized programs as CPM, PERT, PREMIS, PCS and the like. Further, considerable effort had been placed on the utilization of computers for accounting purposes for individual job cost control and budgetary control.

The program pursued was not intended to do away with the aforementioned programs or to refine them or, for that matter, to phase out other more conventional methods of planning and scheduling, such as plan-a-log, bar charts, networks, etc. This in a large part was due to the fact that many of the aforementioned have and will continue to be of inestimable value in the planning and scheduling process. In fact, in most cases, the magnitude or scope of the job in itself helps determine which technique should be employed. Therefore, the intent of the computerized planning and scheduling system was to use existing programs to their maximum in conjunction with a system that would handle day-to-day maintenance activities utilizing resources in the most efficient and effective manner.

In the developmental and implementation phases, planners and field foremen of considerable experience contributed considerable time, numerous ideas and suggestions for the development and implementation of the planning and scheduling system. Many of us during the development phase worked many hours after normal hours to develop the program. This, in large part, was due to the need for developing a program that incorporated maximum input from maintenance personnel or in short developing an "our type program." Use of our input then allowed a programmer from another department in the company to develop a program that would satisfy our requirements.

Equipment selected for the empirically developed system included: a mini-computer employing floppy discs, a printer capable of printing one-hundred and eighty characters per second and a CRT. Input is accomplished thru the use of the CRT (thus eliminating the need for key punch activities) with all planners being capable of operating the CRT for insertion or deletion of information. Discs or diskettes are inserted for particular programs. Printed out information can be obtained in a short period of time if desired by any planner with minor manipulation of the CRT.

The computerized scheduling and planning system is used to maximize resource allocation, permits first line foremen input, assists in control of backlog and permits efficient and timely scheduling. Further, it assists foremen with clerical duties and other paper work enabling them to spend more time in the field, thereby increasing productivity.

I'm sure all of us in the maintenance business have recognized the effects of ineffective planning and scheduling. Further, we all recognize that maintenance contributes nothing to the end value of the product. However, the cost of maintenance is a very significant part of the cost of doing business. Therefore, if we in maintenance can more effectively plan and schedule, dollars can be saved that contribute to the earned profit of the company.

With our computerized system, we schedule for numerous areas including our metal shop, machine shop, garage, projects section (pooled resources) and various zones (allocated resources),etc. Services provided include the development and generation of schedules, normal and capital backlog runs, management information and special requests.

Job work flow includes generation of work in the form of requests, prioritizing of the requests, estimating the work, entering jobs in the backlogs, holding meetings for schedule development, field accomplishment and close out (see Attachment #1).

Schedules are developed for zones utilizing backlogs which are established by inputting daily maintenance requests that are estimated and entered in the computer on the day they are received.

Specific normal maintenance backlog runs include: coding of the requests by year, month, day on which they are received, estimates and other pertinent information required for normal maintenance work. At the refinery, we deal with five to eight hundred maintenance requests per week or thirty to forty thousand maintenance requests per year.

All schedules that are generated include:

1. Safety statistics and safety slogans
2. Coded ticket numbers indicating when the requests were received
3. Scheduling of safety meetings
4. Plant identification
5. Priority
6. Time allowance for P. M.
7. Charge Numbers
8. Jobs that must work
9. Jobs that may be worked at the foremen's discretion (based upon factors such as emergency jobs received during the day, weather conditions, unexpected delays of material, delayed shop work, unanticipated sickness, etc.)
10. In some cases scheduled start and target completion dates

During the last twelve months, we have been able to achieve in excess of seventy-five percent adherence to schedule. The following slides show completed work by all foremen in a given area or zone. You can see from the completed schedules that various write-in jobs are included reflecting emergency or high priority jobs received and worked during the day. Write-in jobs obviously are not considered to be adherence to schedule.

Jobs positioned above the line are considered "must do" work. Those positioned below the line are considered important but the specific jobs to be worked are determined by the first line foreman. Any jobs below the line which are not started on the date or day assigned are not worked the following day but reconsidered for accomplishment at the next weekly scheduling meeting. Schedules are developed based upon 120 percent of possible work (50% above/ 70% below) on Tuesdays thru Fridays and 100 percent (35% above/65% below) possible work for Mondays. The reason for the Monday percentage is due to the number of requests normally received following the weekend. It should be recognized that a 75 percent adherence to schedule in reality amounts to 87 percent completion of the schedule i.e., (75 X 100) + (75 X 120 X 4) ÷ 5 = .87%.

In addition to schedules and backlogs in the zone programs, management reports can be requested. Management reports show the number of requests outstanding and the corresponding total number of manweeks required to accomplish the backlog in a priority listing the number of requests received in a given period and the number of steam leaks outstanding in a given zone.

Special request or computer backlog runs can also be generated based on descriptors coded in the backlog (See attachment #2). As an example, if it is desired to produce a listing of outstanding electrical jobs, only electrical jobs would be generated on the run.

For the Zone portion of our programs, the following items can be accomplished via the CRT: (See Attachment # 3)

1.   Load data into zone files
2.   Prepare Maintenance Backlog report
3.   Display work orders by Supervisor and by Plant
4.   Display work orders by Supervisor and by priority
5.   Calculate the number of jobs and hours or weeks by priority for a total zone.
6.   Calculate the number of jobs and hours or weeks by priority by foreman.
7.   Close out finished work
8.   Backup data files
9.   Prepare Maintenance Schedule
10.   Change content of data files
11.   Print backlog by Area Supervisor and or specific plant (See Attachment #4)
12.   Print Backlog by Area Supervisor after a specified date.
13.   Delete finished work from data files
14.   Print special backlog reports by type of work

Another program incorporated in our planning scheduling system is the Capital program. Our Capital program includes jobs that require engineering effort for accomplishment, including jobs that are truly capital in nature (plant impovement) or large expense jobs requiring engineering effort for accomplishment. Information found in the program includes the specific planner responsible for planning the job, date the engineering job was received, work schedule commencement date, anticipated completion date, percentage of field completion, etc.

Shown is a typical page of a Capital Backlog run. As with the zone program, schedules can be generated, backlogs run and special requests printed based upon the use of descriptors.

When engineering packages are received, planners review the

packages for needed materials, potential problems, safety requirements, special equipment , major tools needed, etc. and field checks the area. The planner then selects the appropriate method to be used for detailed planning of the job. This might include use of the plan-a-log, networking, bar charting or other established planning methods. During this phase, particular attention is paid to incorporating safety criteria into the job, recognizing budget and time constraints, etc. Further, the planner manloads, allocates resources, sequences activities, develops a detailed schedule and strategy for field accomplishment and budget monitoring. Armed with this information, he assigns the work to the appropriate area for field accomplishment.

Further expansion of our computerized scheduling is the distribution of mobile equipment in the refinery which is accomplished by the generation of a weekly schedule. This program provides a permanent record of what equipment is assigned to specific jobs and also a further breakdown of charges incurred on different types of jobs-shutdown, expense, capital. Monitoring of budget and actual data can be accomplished on a daily basis if desired.

Work in our garage is scheduled daily. This allows for flexibility in providing manpower to accomplish repairs on unexpected equipment breakdowns. State inspections and PM work on refinery vehicles are appropriately scheduled.

An additional area now using computerized scheduling is our Shops which consists of Metals, Machine, and Builders. Although very similar to the Zone program, there is added flexibility. Because most work received by these shops is from other geographical locations, the Shops program has the capability of printing a backlog for each of these areas. Comparison of estimated and actual hours appears on the backlog, along with projected starting and completion dates for each job.

An important benefit derived from computerized scheduling is a steady decrease in normal backlog in all areas. Better control of work in all areas is another important result achieved.

A PREVENTIVE MAINTENANCE PROGRAM

Another computerized program utilized at Sun's Marcus Hook Refinery is a total PM program. PM makes sense and is used extensively by all of us in our daily lives. Consider for a moment the amount of PM performed on your automobile in a given year: oil changes, air filter changes, greasing, washing, waxing, adding water, tune-ups and so on. Or one might conclude that we even perform PM on our bodies when we have an eye exam or physical. It therefore seems logical that total PM can be applied to a refinery-- since we all recognize our equipment is working for us twenty-four hours a day and deserves some attention. The degree of attention becomes the question. At our refinery, we feel to some extent we have been successful answering this question. Let's consider the requirements necessary for developing and implementing an effective PM program:

1. Clearly defined responsibility and authority
   A. Determination of equipment - Op/Maint.
   B. Ops. Involvement-Active input from Ops.
2. Proper Organization
3. Adequate procedures and control
4. Proper operation of equipment
5. Ops/Maint carry out PM functions
6. Good cooperation necessary between Ops and Maintenance which should be a common goal
7. Maint-Inspect/Lube/Repair, etc.
8. Update Maintenance equipment history
9. Downtime figures
10. Realistic Cost Data
11. Must be supported by all levels of management
12. Planning/Scheduling
13. Economical justification for PM
14. Flexibility of PM system
15. Cooperation and understanding

Advantages and fringe benefits received thru the use of a PM program include:

1. Minimal paper
2. Pre-scheduled assignments
3. Load balanced workload
4. Flexibility for change
5. Ease of issue/scheduling
6. Ease of control
7. Minimal record keeping
8. Spreading of load
9. Suppressed items

Fringe Benefits

1. Allows us to work smarter, not harder
2. Promotes pride
3. Enhances team work/cooperation
4. Promotes better working conditions
5. Enhances safety awareness
6. Promotes smoother functioning
7. Provides a tool for doing our jobs more efficiently with minimal output
8. Assignment of small jobs (PM activites will help eliminate early arrival in Shops)

Now that we have addressed the requirements, advantages and fringes of a PM program, we should also consider perhaps the most important fact of all when developing and implementing an effective PM program; namely, "PM programs don't just happen." It takes a lot of hard work, knowledge, skill, experience background, planning, organizing and tenacity or "can do spirit" plus management support at all levels to develop and implement an effective PM program.

Planning and organizing for a PM program includes:
A. Defining Preventive Maintenance
B. Establishing objectives
C. Determination of equipment to be included in program
D. Obtaining equipment performance data
E. Estimating costs
F. Allocation of personnel (recommending)
G. Basic tools
H. How much Preventive Maintenance is enough?

The philosophy for development is easy--keep it simple, minimize paper work, and ensure that all parties involved recognize that PM is everybody's responsibility. Obviously goals, objectives and potential returns, and planning strategy should be considered prior to attempting to develop a total PM program.

An initial effort to devise a PM program resulted in the generation of computer runs. These were produced weekly on large sheets of computer paper. The program had been formulated with minimal input from field foremen and, as a result, received only token acceptance. Our intent, then, was to make use of the vast information contained in this program, but to modify it to better meet our refinery's needs and make it a viable tool in the Maintenance Department.

In our case, staffing for the PM program development and implementation included myself and three experienced foremen that were selected based upon their knowledge of the overall refinery, ability to interface effectively with others, their"can do spirit," ability to adapt to change, etc. The aforementioned were placed on the development and implementation full time. Further for the instrument PM activities, various key personnel devoted part time on the project.

We first defined PM. Simple-right? Wrong! We identified in excess of one hundred definitions of PM and expanded on that number when we asked various individuals for their definition. The definition we use is as follows:

That type of Maintenance which inspects, adjusts, etc. equipment at regular intervals before failure of any one of these individual units could force an emergency process shutdown or in short, pay me now or pay me later.

After that, we were on our way. We inputted into the program the following:

1. Standardization of PM activities
2. Manpower optimization
3. Planning, Scheduling techniques
4. PM Paths
5. Experience
6. Operations input
7. Maintenance input
8. Critical equipment defined

and have achieved output or results such as:

1. Identification of job responsibility
2. Operations PM activities
3. Maintenance PM activities
4. Extended equipment life
5. Safety awareness
6. Reduction in breakdowns
7. Optimum timing and flexibility
8. Manpower requirement defined
9. Lubricants defined
10. Minimized paper work
11. Better housekeeping

Our team analyzed individual pieces of equipment thoughout the refinery determining if in fact PM was warranted, precisely what PM activities should be performed, methods to utilize,and at what frequency PM activities should be performed. Equipment covered and methods utilized in the PM program include:

1. Any equipment that economically warrants PM coverage.

    a. Mechanical/rotating equipment
    b. Electrical equipment
    c. Instrumentation
    d. Safety
    e. Energy conservation

2. Examples of Zone mechanical equipment being covered under
    the PM program.

    a. Towers (walkways and ladders)
    b. Accumulators (safeties and valves)
    c. Platforms (structural strength, etc.)
    d. Safety items (handrails, etc.)
    e. Pumps (couplings, seals, etc.)
    f. Motors (Lubrication, efficiency, etc.)
    g. Exchangers (pressure/temp.)
    h. Reboilers (efficiency)
    i. Lighting (safety, operations)
    j. Housekeeping (cleanliness)

3. Examples of instrument equipment being covered under the
    PM program.

    a. Energy conservation equipment
    b. O_2 analyzers
    c. Hydrocarbon detectors

Typical PM activities for mechanics include:

    1. Alignment
    2. Adjusting
    3. Cleaning
    4. Checking
    5. Change and flushing
    6. Safety items
    7. Housekeeping
    8. Lubrication

Instrument PM activities include:

1. Checks
2. Adjustments
3. Calibrations
4. Correlate
5. Diagnosis
6. Reaming
7. Timing
8. Recording

Typical operator PM activities include:

1. Check steam traps
2. Check oil levels
3. Grease yoke **nuts**
4. Check seals
5. Check safety items
6. Housekeeping
7. Check/Note leaks
8. Check lighting

* Where zone maintenance is required item is to be noted on reverse side of card.

In the case of this program, input to the computer was accomplished via keypunch. Input information included:

1. Location codes
2. Parts defined
3. Methods defined
4. Responsibility defined
5. Lubricants listed, where applicable
6. Special instructions noted
7. Frequency defined
8. Suppress codes utilized

Information fed into the computer was manloaded by the computer over a twelve-month period. This helps ensure that any given area would receive approximately the same amount of work each week during the twelve-month period. Further some information was fed into the computer for specific dates. As an example, turning on of warming lines by operators was fed in for November.

Output of PM activities is via cards about the size of a pay check. Information contained on the cards includes:
Location, who job is assigned to, equipment identification and description of PM required, and frequency of PM required.

The cycle used for issuance thru card return includes:
Cards generated from the computer are first issued to the appropriate zone area. It is then issued to the Zone Foreman and Operations. The foreman issues it to the Zone mechanic and Oper. Foreman. After completion of the work, the mechanic returns the PM card to the Zone foreman and Oper. Office. It is returned to the PM coordinator for verification.

The question of where we get the manpower from is as follows:
1. Historically, it has been proven that when effective PM was employed, the overall workload decreased.

2. You can anticipate that initially it will be tough and backlogs may increase slightly. However, within a reasonable period of time, if you faithfully adhere to the system, backlogs will decrease and overall operational costs will be reduced.

Management control of the PM program is accomplished through:

1. Monitoring
2. Random sampling
3. Weekly cost system follow up
4. Radnor system
5. Good management

Specific examples of the effectiveness of this new PM program appeared in the refinery's weekly paper. Gear cases containing no lubrication were found. Shutdown time was avoided by discovering a worn sprocket on an elevator and replacing it on off shifts. Cooling tower motors were taken out of service one at a time and realigned to avoid a costly emergency failure at a later more critical time.

Results achieved were considerably better than projected. We had initially projected that pay back would be achieved within one year. Pay back in reality was within six months. The first week the program was in effect the computer costs for the entire year were paid.

These are two of the many computer programs utilized at Sun's Marcus Hook refinery, enabling us to better manage all of our resources.

Mr. Leone is currently the Manager of Maintenance Systems for Auto Tell Services

ATTACHMENT 1

MAINTENANCE PLANNING
JOB WORK FLOW

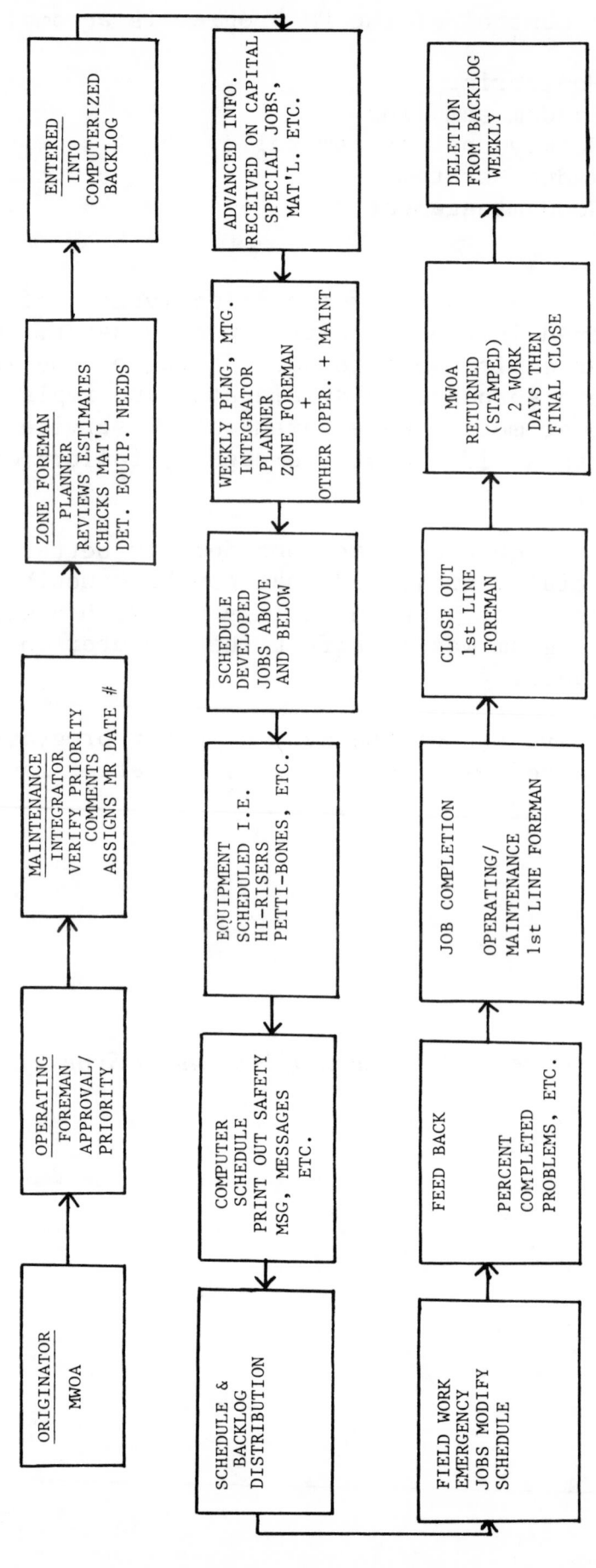

DESCRIPTORS

SI - Safety
** - Leones
SD - S/D Items
PO - Pollution
DI - Dismantling
RT - Retube
CT - Contract
PC - Purchase
AC - Awaiting Close Out
EL - Electrical
PA - Painting
SC - Scaffolds
IS - Insulation
$$ - Leones
PU - Pump
IN - Instrument
VA - Valves
MW - Mechanical
CC - Capital Construction
UA - Unassigned
CA - Capital (70's)
MC - Minor Capital (30's-40's-50's)
SL - Steam Leaks
LA - Labor
CO - Compressors
BM - Building Maintenance

MENU

A. Load data into Zone files.

B. Prepare Maintenance Backlog report.

C. Display Work Orders by Supervisor and by plant.

D. Display Work Orders by Supervisor and by priority.

E. Calculate # of jobs and hours/weeks by priority-total Zone.

F. Calculate # of jobs and hours/weeks by priority-by Area Supervisor.

G. Close out finished work.

H. Backup data files.

I. Prepare Maintenance Schedules.

J. Change content of Data files.

K. Print backlog by Area Supervisor/plant.

L. Print backlog by Area Supervisor and after MR#.

M. Delete finished work from data files.

N. Special backlog reports-by type.

ATTACHMENT FOUR

ZONE 5

SAFETY MEETINGS

TUESDAY 7:45 AM	FRIDAY 7:45 AM	THURSDAY 7:45 AM	WEDNESDAY 7:45 AM	MONDAY 7:45 AM	THURSDAY 7:45 AM
CASSIDY	ALLEN	CAPEL	BALENTINE	BURLEW	MC ABEE
AREA A1	AREA A2	AREA A3	AREA A4	AREA A5	AREA 6
ALKY B1114	POLY	17-1A B1146	17-1FP B1147	C O M P R E S S O R S	
15-2A	15-1 B1116	17-P B1150	17-1H B1149		
15-2B 2	15-5 B1074	17-2A B1155	17-2 B1151		
15-2A B1073	BH B1221 B1232	17-3 B1154	17-TA		
15-2R	WP B1233 B1234	6-SUB B1222	17-TR		
SWEET B1120	TA	RWC B1152	17-CR		
RW-7 B1321	4-SUB B1125	ZONE B1367	17-GEN B1148		
15 LAB	15-CR				
15 GEN					

237

Reprinted from *Power*, June 1980

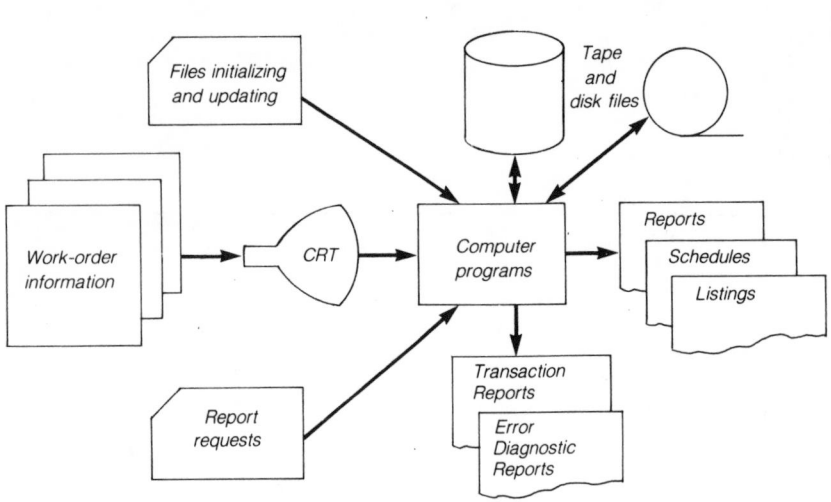

1. Computer-based WOSS© generates work schedules, priorities, variety of reports

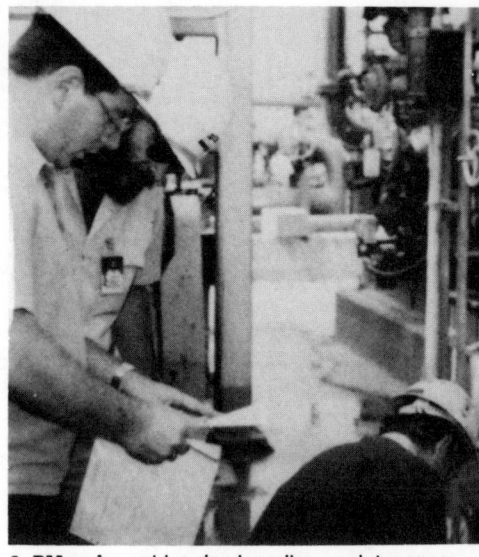

2. PM order guides demineralizer maintenance

Energy management

Managing and Controlling Maintenance

A well-planned maintenance-management program can bring organized
yet flexible scheduling, improved productivity, and a ready-access
data bank. The system detailed here, developed for a nuclear
power plant, has application to any equipment-intensive facility

By G Chavez, Southern California Edison Co, and **H F Perla,** Pickard, Lowe & Garrick Inc, Irvine, Calif

During construction of San Onofre generation Station Units 2 and 3, early preparations were made to assure that a sound maintenance-control system would be available for plant startup, and applicable to normal operations thereafter. Effective maintenance management can provide information that is vital for productivity improvement, at the same time that it controls day-to-day maintenance activities; thus, the timing of the project presented a golden opportunity to design a system responsive to both needs. A similar program could do the same for other plants, regardless of type.

Consistent but flexible control

Southern California Edison's experience at its other generating units gave us the basis for establishing the system criteria. We supplemented this with visits to other large electric utilities, to review their experience and maintenance-management approaches. With this background, we determined that the end-product would provide three invaluable tools: maintenance control, a preventive-maintenance (PM) program, and historical records that are readily accessible.

The maintenance control requirement was to provide a mechanism to assist the maintenance-planning staff in setting priorities and scheduling work in a consistent manner, based on the job's importance, and regardless of the origin of the request. The control was to provide flexibility in scheduling and reports so that major types of maintenance work—preventive or corrective maintenance, retrofits, or overhauls—could be separated or integrated, as desired. It was also to permit integration or separation of reports by one or more assigned groups—such as electrical, mechanical, or instrumentation and control—so flexibility in both organizational responsibility and supervision structure could be accommodated.

The PM control program also had several requirements. It had to readily identify when individual preventive maintenance, surveillance, test, and calibration activities were due, and had to establish the importance of each activity in a way consistent with the other maintenance activities that were to be performed. The program had to be automated sufficiently to minimize—or even completely eliminate—the need to reinitiate a work order each time a PM or other repetitive activity was required. It had to accommodate the ganging of multiple components into a single repetitive work order, so the amount of paper controls required for similar components could be minimized without sacrificing control. Further, the PM control program had to provide some feedback so that the frequency could be increased or decreased if surveillance, test, and calibration activities were not yielding results commensurate with the effort being applied.

Historical information, the third requirement, was tied to the need to support a productivity-improvement program, which requires certain equipment and system-performance information. The maintenance-management system had to permit reasonably fast retrieval of work histories, regardless of type, for each component or system in the plant. Further, it had to accumulate the failure histories of systems and components,

```
REPORT - WSB610P03              SOUTHERN CALIFORNIA EDISON
                            SAN ONOFRE NUCLEAR GENERATING STATION
                                   WORK SCHEDULE REPORT
                                        02/01/80

      ASSIGNMENT - STATION MECHANICAL/ELECTRICAL
      MAINTENANCE - PREVENTIVE AND CORRECTIVE MAINTENANCE

                                            HOURS
      --------------W O R K  O R D E R / P M /-----------  TO   QUALITY PLANT ESTIM      LABOR      ESTIM
      PRTY NUMBER        DESCRIPTION          DATE   IMPACT CLASS  COND  DUR  LATE CLASS  SEQ   M/H

        8  WO000017  FIRE WTR PP ELECTRIC DRIVEN 01/02/80    2     A     7         ELEC   2    1.0

                                                                                   LAB    1    5.0

                                                                                   MW     3    5.2


        9  PM300017                          06/25/80       1           3         ELEC
        9  PM100012                          12/25/80       3           1         ELEC
       10  WO000010  AIR COMPRESSOR          01/02/60       3     C     4         ELEC   1    4.0
       10  WO000015  480V DISC SW ADO29      01/02/80 ·     3     D    16         ELEC   1,3  3.0

                                                                                   PF     2    8.5

       10  WO000019  INTAKE STRCUTE SUM PP N 01/02/80       3     A     1         MW     1    1.0
```

3. Work Schedule Report lists work orders (WO) and PM orders in descending priority, is basis for work assignments. 'Hours to Impact' applies to nuclear-plant specs

particularly with respect to resulting loss of plant power.

Against these criteria—supported by Control Data Corp's Professional Services Div—we designed a Work Order Scheduling System—trademarked WOSS—as a maintenance-management system for use on SoCal Edison's IBM computer system. WOSS tracks all work orders, from the time they are initiated until they have been completed, then stores selected historical information (Fig 1). Key features of the system include:

■ Documentation of all work requirements and their completion.

■ Automated prioritizing of work orders.

■ Automatic generation of work orders for repetitive work, when due.

■ Work lists in priority sequence.

■ Deferred, standby, and outage work reports.

■ Labor and work-order backlog information.

■ Equipment work histories and cumulative failure data.

The system provides outputs in the form of reports for use by several levels of personnel, from craftsmen to managers. Work schedule, deferral, standby, outage, and backlog reports can be separated according to maintenance type and responsible organization sector. Multiple components can be ganged into single repetitive work orders for preventive maintenance, surveillance, test, and calibration activities.

The WOSS was designed to be used routinely by the maintenance planner at the plant where the work-assignment process is located and where the system must be most responsive. It comprises 19 programs, each performing specific functions that use various computerized ref-

erence files. Thus, the system is modularized for efficiency, so only the functions to be performed are executed, without requiring the computer to operate the entire system.

There are two types of files in WOSS: Those that are dynamic, where information is frequently changed; and those that are relatively static, and which, once initialized, are modified infrequently. Most of the files are designed for disk storage, although tape storage is used for two history files of constantly increasing size.

Cathode-ray tube (CRT) remote job-entry terminals at the plant are used to update the work-order files, since these must be modified each shift, or at least each day. Keypunched cards and a card reader—or CRT, if the job is small enough—are used to initialize and update other relatively static files, and to request reports, schedules, and file listings. Both input processes communicate with a computer located at the general office. In addition to the working reports, schedules, and listings, the computer produces transaction reports and error-diagnostic reports after each program has been processed, thereby providing an audit trail of the actions taken.

The computer automatically sets priorities on work orders, based on a combination of work-order information, supplied by the maintenance planner through the CRT terminal, and system and component information already stored in the computerized files. The priority sequence, in descending order of importance, is based on:

■ Emergency nature of the work (personnel and plant hazards).

■ Power-production limitation experienced.

■ Regulations and commitments made in the technical specifications.

■ Schedule constraints.

■ Potential hazard if the work is not accomplished.

■ Component or system criticality to plant production if it should fail.

■ Other legal constraints.

■ All other work.

By using the work-order information, which indicates the emergency or potential emergency nature of the problem and the loss in plant capacity or productivity, if any, and accessing the reference files containing the other priority information, the program's logic assigns the highest qualifying priority to the work order. It then ranks the work order with those already in each priority group, on the basis of occurrence date or other criteria established for the group.

All reports indicating work orders ready for implementation list them in priority sequence; this reflects the order in which work should generally be done. However, it is the maintenance planner who makes the actual work assignments. He can decide that a lower-priority work order should be executed earlier, if he has special knowledge indicating such action is warranted.

Reference files form data base

The WOSS contains files that provide logical data separation and structure, to enable the programs to function efficiently. Each individual program calls on one or more of these files. The largest is the System/Component Master File, which contains a record for each plant component. Included in each such record are its identification number (ID) and description, system ID and unit number, and component quality class. Also included for maintainable components are any applicable PM procedure numbers and their frequency, and, for each combination of these, the required plant and equipment operating condition, the labor crafts required, assignment responsibility, and duration. File records for equipment covered by a surveillance, test, or calibration activity include similar information.

The records in the System/Component Master File also contain the information used in setting priorities on work orders: A "critical-to-operations" ranking based on plant capacity lost if the component fails; an availability ranking based on failure frequencies experienced; a legal-obligation indicator; and the date of scheduled need. Cumulative failure data in the records include the number of component failures, hours out of service, and megawatt-hours lost due to the failures.

Since San Onofre is a nuclear plant, there is a Technical Specification Master File, which relates plant-system IDs with

```
REPORT - WSB610P02                   SOUTHERN CALIFORNIA EDISON
                               SAN ONOFRE NUCLEAR GENERATING STATION
                                          BACKLOG REPORT
                                            02/01/80

    ASSIGNMENT - TOTAL ASSIGNMENTS
    MAINTENANCE - TOTAL MAINTENANCE
       LABOR
    CLASSIFICATION              READY WORK    DEFERRED     SUBTOTAL      OUTAGE       TOTAL

    ADMINISTRATORS                             100.5        100.5                    100.5
    BOILERMAKERS                    9.0          1.2         10.2         13.6        23.8
    CARPENTERS                      6.0          3.4          9.4          1.0        10.4
    ELECTRICIANS                   15.5          2.8         18.3         19.9        38.2
    INSULATORS                                  92.3         92.3                     92.3
    INSTRUMENT TECHS               10.0          2.0         12.0          3.0        15.0
    LABORERS                        5.0         12.0         17.0                     17.0
    MILLWRIGHTS                    19.2          1.0         20.2          8.0        28.2
    OPERATING ENGINEERS             8.0         23.0         31.0                     31.0
    OPERATORS                                    1.5          1.5                      1.5
    OTHER                           4.0          4.5          8.5                      8.5
    PIPEFITTERS                    12.0         12.5         24.5          3.0        27.5
    PAINTERS                                     2.5          2.5                      2.5
    SHEET METAL WORKERS             6.0         12.3         18.3          2.5        20.8
    TEST TECHNICIANS                1.0           .5          1.5                      1.5
                               ---------    ---------    ---------    ---------    ---------
       TOTAL LABOR MANHOURS        95.7        272.0        367.7         51.0       418.7

       WORK ORDER COUNT             16            2           18            7          25
       PM ORDER COUNT               59            3           62            5          67
                               ---------    ---------    ---------    ---------    ---------
       TOTAL ORDER COUNT            75            5           80           12          92
```

4. Backlog Report summarizes ready, deferred, and outage work in terms of manhours, by labor craft. Periodic plot reveals trend in work buildup or reduction

sections of the Tech Spec, and is also used in setting priorities for work orders. It indicates the maximum number of hours after discovery of a deficiency before an impact is to be imposed.

PM-Order Master Files retain a record for each repetitive preventive-maintenance, surveillance, test, and calibration activity. These files permit aggregation of like components that have the same frequency and use the same procedures. For example, a number of valves or instruments can be grouped into a single repetitive work order, without having to generate individual work orders for each. Each file record holds the information common to all the components in a PM order, and references each component to which it is applied (Fig 2).

The WOSS also contains a Work Order Master File. More dynamic than most of the other files, it contains a record for each active work order that has been issued but is not yet completed or canceled. Consequently, the file's contents are constantly changing. This file is the prime source of information for many of the system reports.

Other files included in WOSS are an Equipment Work History File, a PM-Frequency History File, and a PM-Procedure Description File. These are used to support the various parts of WOSS discussed below.

PM orders for repetitive work

The function of the programs associated with this part of WOSS is to generate automatically work orders that are used to implement periodic preventive maintenance, surveillance, test, or calibration work when due. These work orders, or PM orders, are numbered by the computer and grouped according to assigned discipline.

When work is planned for a given period—the coming week, say—the pro-

gram scans the PM Order Master File and identifies each PM task whose next due date falls during that period, and for which an order has not already been issued. The PM orders produced by the computer include this information:

- Priority number.
- Reference procedure numbers, titles, and frequency.
- Component IDs and titles covered by the PM order, their quality class, and the operating condition required for the work.
- Labor classifications (crafts) required and average hours used in the past by each.
- Estimated duration of the job.
- A special notification, if any exception was encountered in a prior execution.
- Verification-point approvals for equipment relating to Quality Class 1 or 2, ASME Code, or fire-protection systems.
- Signature blocks showing acceptable PM order completion.

A PM and S/T&C Due Report (latter denotes surveillance, test, and calibration) is also issued, listing the PM orders just issued by the computer. It includes information enabling supervisory monitoring of PM completions.

As each PM order is completed and closed out, the computer calculates the next scheduled due date, based on the completion date and the specified frequency, and logs in the next due date in the PM Order Master File. The same program increases the record of the number of PM order executions by one, and recalculates a new average of labor hours used by each craft in performing the work. In addition, completion information is added to the PM Frequency History File and Equipment History File. The PM order is then removed from the Work Order Master File, which con-

tains all active PM orders and work orders.

A record, maintained in the PM Frequency History File for each PM order, indicates whether each component covered by the PM order was found to be within acceptable tolerance. On request, by PM order number, a PM Frequency History Report is produced, showing all executions of the PM order and the tolerance acceptability. The report is used to evaluate the suitability of either maintaining or changing the frequency assigned to that PM order.

Sequencing and deferring work

The work-order control system is based on the use of (1) work orders, to document the work required, and (2) work-order reports, from which work is scheduled and controlled. The work-order form is designed primarily to provide information needed by maintenance personnel to plan and accomplish the work. However, it also contains information used by the WOSS to establish the work order's priority and to categorize it, so it can be tracked from initiation until completion. The WOSS information included on the form includes:

- Unit number, and one or more system or component IDs.
- Indication if an emergency involves personnel safety or plant hazard.
- Percent of plant capacity lost.
- Tech Spec application.
- Crafts, sequence, and estimated hours for each.
- Job duration.
- Required plant condition.
- Deferral reason and responsibility for resolving.
- Maintenance type and assignment.

Using the information from both the work order and the computer files, reports and schedules are generated on request. The reports are keyed to selected maintenance type (PM, CM, retrofit, etc) and to the assigned staff group (electrical, mechanical, I&C, etc), or to combinations of maintenance types and assigned groups. The variety of available reports are described below.

The Work Schedule Report (Fig 3) integrates all work orders and PM orders (unless selected as separate maintenance-type reports) for which all required tools and materials are available, provided the orders are ready for implementation as soon as labor is available. The work orders and PM orders are listed on the report in descending priority and discovery date. This report becomes the basis for work assignments, with the work orders at the top of the list generally accomplished first. It contains:

- Work-order or PM-order description.
- Tech Spec hours before impact is imposed.

directly to production as compared with 31 percent for those that do not employ mathematical models. This was explained as follows: (1) maintenance units with high status have greater responsibilities, and (2) low status maintenance units are much more limited in their policies.

Turban (17) also found that the existence of OR units at higher (divisional or corporate) levels is a factor positively correlated with the use of mathematical maintenance models (Group A - 78%; Group B - 55%, a significant difference at .025 level).** The existence of maintenance units at division or corporate level is also a factor positively associated with the use of mathematical models (Group A - 58%; Group B - 42%, a significant difference at .10 level). Existence of OR and maintenance units located at higher hierarchical levels facilitates implementation and utilization of mathematical models by the ability to conduct comparative studies, better participation and communication across industries, departments and professions, better and larger bases, possibly higher level of experience and education.

A very important element, which without any doubt is overlooked by the researchers, is the organizational structure of maintenance. Failure to account for this element hinders the development of practically applicable models (e.g., optimizing repair crew sizes) and consequent implementation of these models in industry. Turban's study (17) also does not provide us with clues as to how maintenance organizational structure is correlated with the usage of mathematical maintenance models. It seems to us that any future empirical study of this nature has to investigate the relationship between organizational structure and maintenance models and therefore it is worthwhile to examine the major aspects of this problem.

There is no single universal model for the ideal organizational structure of the maintenance function. The structure and resources should be flexible to the needs of a specific plant. Basically a good maintenance organization may be organized by using:

1. Decentralized (known also as area) system
2. Centralized (known also as functional) maintenance system
3. Combined maintenance system

In a decentralized maintenance organization, the members of a repair crew are assigned to specific areas in the plant. The areas may be defined geographically, by product or production grouping and/or by service function (utilities, measurement laboratory, etc.).

In a centralized maintenance organization maintenance mechanics and craftsmen are assigned to work at any or all areas of the plant. The number of repairmen may vary from one maintenance worker in a very small organization to 1000 employees in such large production organizations as oil refineries.

** Group A includes plants which use mathematical and operations research models in maintenance decision making; Group B includes plants which do not use quantitative maintenance models.

The controversy about centralization/ decentralization still continues to be debated by the maintenance specialists. Usually a decision as to what system to use is based on cost-benefit analysis. The benefits associated with the decentralized maintenance system are realized in speed of response to a call for maintenance and the close familiarity of the repair men with the equipment. The decentralized workplace will be likely to diagnose faults more quickly than maintenance specialists from a central maintenance workshop, who are required to deal with the entire range of operational equipment. When decentralized organization of maintenance is used there should be less need for an elaborate documentation and communication system.

However, in a decentralized maintenance system it is often difficult to achieve high utilization of personnel and very expensive repair equipment and tools. There is a much better potential for a higher resource utilization in a centralized maintenance system. In addition, the cost of managing can be excessive if a foreman is required for each of a number of decentralized groups.

Arguments against the centralized maintenance usually come from production managers, who are accustomed to work in an ever-present environment of frequent and unpredictable plant stoppages due to emergency breakdowns. As a result of this, they quite understandably insist on having maintenance men at their immediate request.

Many plants attempt to resolve the problem of balancing service and maintenance costs by combining a central workshop (maintenance department) with an area or departmental maintenance group. The variations of such combinations are unlimited. The advantages and disadvantages of the basic system exist in proportion to their contribution to any one combined system. This combined system is also known as the "cascade" decentralized/centralized system of maintenance organization.

In this scheme decentralized maintenance groups are helped by the central maintenance department, which is equipped with skillful repairmen and a wide range of required equipment. When peak workloads are encountered they are allowed to cascade or spill over from the local workshops to be met by the central maintenance department. In case the central maintenance department is overloaded the area maintenance work force picks up the excessive work volume. When excessive or highly specialized workloads are accumulated by the central department and in-plant personnel is not able to handle the excess load, it is allowed to cascade or spill over to outside maintenance contractors.

There is no empirical evidence about the relationship between usage of operations research models and contract maintenance. An attempt to incorporate contract maintenance in a mathematical model dealing with determination of optimal size of a maintenance work force to meet a fluctuating workload was made by A. Jardine (7). However, no reports of application of this or similar models were found in the literature.

The use of contract maintenance has increased considerably in recent years, though at one time it was seen to be of value only in domestic maintenance (building repairs, roofing, pipe installation, etc.). However, a growing number of firms now make use of contractors for the maintenance of standard production equipment.

Many maintenance managers assert that jobs can be done more cheaply by outside contractors than by their own maintenance employees. Others, equally emphatic, state that there is often no place for contractors to assist with maintenance, and only new installations warrant their engagement.

The decision to provide the necessary maintenance personnel and required machinery within the firm's organization (in-house maintenance) or to utilize services external to the organization (contract maintenance) is primarily an economic decision. The problem arises in evaluating the economics of alternatives.

Three major alternatives are available today to maintenance managers:

1. In-house maintenance - all maintenance work is done by the internal maintenance work force.
2. Combined in-house/contract maintenance - also known as "peak shaving" maintenance, which simply means that under normal periods of plant operation the general maintenance load is handled by a small existing staff. However, during periods of shutdowns and major equipment overhauls the extraordinary work will be subcontracted to an independent service organization.
3. Full contract maintenance - under this type of maintenance program an outside maintenance contractor is hired to plan, supervise and handle entire maintenance operations, providing all requirement management, foremen, personnel, equipment and supplies.

J. Jordan (8) determined that there are four key areas that management should explore before deciding whether or not to use contract maintenance: (1) labor relations, (2) comparative costs, (3) quality of work, and (4) flexibility.

Definite conclusions concerning labor relations were these:

1. There are no serious labor relations problems.
2. Contract maintenance will not cause unionization of a plant.
3. There have been very few work stoppages under contract maintenance.
4. It is easier to get rid of poor workers.
5. In-plant unions will not prevent good contract maintenance.

The findings on comparative costs were inconsistent. However, the survey respondents largely agreed that there is very little "make work" in contract maintenance and only 28 percent felt that contractors' profit was too high.

Conclusions on work quality under maintenance were as follows:

1. Contract maintenance offers an opportunity to secure a work force of good quality.
2. Effectiveness depends on the quality of supervision.

Jordan found that the biggest advantage of contract maintenance is flexibility. The ability to recruit and dissolve a large work force quickly was most often cited as the major advantage of contract maintenance. Other advantages cited by Jordan were (1) contract maintenance is applicable to more than turnaround work, and (2) it is hard to cut down in-plant maintenance force.

One case of high flexibility was reported by Rohrmann (14). The author found that general maintenance contractors' manpower has fluctuated between 110 and 924 men, based on the needs of Getty Oil Company in Delaware City, Delaware.

In Eastern European countries, especially in the USSR, contract maintenance is known as inter-plant maintenance centralization-- the concept and practical application of which have been actively promulgated from the late 1960's. The purpose of large centralized maintenance plants is to transfer repair operations (usually overhaul maintenance) from unit type to batch type of process and therefore to develop and apply industrial methods of maintenance. There are many books, papers and reports on economic efficiency of centralized maintenance, including those of Gnitecki and Volkov (4), Paulukonis, Staroselsky and Chanin (12) and Konson (10). However, insufficient management, administrative red tape, poor state of centralized spare parts production, and lack of financial resources hinders further development of centralized maintenance in the Soviet Union.

Examination of organizational determinants of maintenance management shows that as maintenance operations become more centrally oriented (from area to central maintenance department, from in-house to contract maintenance (and any combination) it requires more coordination, thoughtful planning and scheduling). However, this development also creates large potentials for a wide implementation and utilization of comprehensive mathematical and operations research models in maintenance management.

CONCLUSIONS

Our examination of the literature in the area of maintenance management, published in the last 25 years in the United States, Soviet Union, Great Britain and West Germany, evidences that there exists an extremely wide gap between theory and practice of maintenance. Theoreticians and practitioners of maintenance continue to speak in "different languages" and perceive different interests. The existing empirical studies show that application of mathematical methods and OR techniques is at the lowest level as compared to the other areas of industrial activity. There is almost no information about specific mathematical models or OR techniques utilization in particular areas of maintenance management. There is a large number of maintenance models in the literature based on mathematical optimization and OR techniques, as a result of different assumptions about the state of technical systems, time-to-failure distributions, maintenance policies, etc. It should also be noted that very few, if any, practical applications of theoretical maintenance models are reported. It is also found that many assumptions of these models are often not warranted.

Interpreting the low use of OR techniques in maintenance, Hovey and Wagner (5) and Turban (17) have concluded that this does not indicate low potential for OR application in maintenance, but that it is a broad, untapped area for research.

Agreeing with this statement, we have to note that one of the reasons for the low utilization of mathematical and OR models is the problem of application of these models to maintenance practice. An analysis of the maintenance models, reported in the literature, brings us to the following conclusions:

1. Most of the models are dealing with deterioration, inspection, replacement and repair problems of one-component (unit) or one-machine systems, which, from a practical point of view, have little utility;

2. There exists an over-emphasis on replacement and/or inspection policies. Very often repair operations are not considered or they are given a simplified role in maintenance management models;

3. Most of the models and studies deal with stochastic nature of failure distribution, but very few with stochastic repair operations;

4. There are only a few models or studies of multi-part, multi-echelon, and multi-machine interaction maintenance problems. However, most of these models also assume homogeneous (identical equipment) populations; and

5. There are many behavioral problems associated with implementing theoretical maintenance models. Maintenance managers feel threatened by the complexity of these models and resist changes that the implementation of these models may bring.

As a result of existing deficiencies, the cost of maintaining equipment in good working condition is very high and maintenance operations exhibit very low productivity and low efficiency. The organization of maintenance departments and operations which differ from one plant to another is not adequate. The status of maintenance departments and managers is generally low. There is almost no conclusive evidence of the relationship between implementation and utilization of advanced maintenance models and organizational structure (decentralized/centralized/"cascade").

In order for theory of maintenance to advance and facilitate a cardinal improvement of the management practice, it is necessary that the current state-of-the-art in maintenance be looked at critically and integrated. The task of future research is to analyze and develop premises that will help in bridging the enormous gap between theory and practice of maintenance. The future research may go into various directions and touch upon various problems and issues of maintenance management. One direction is to examine the empirical laws, which govern the current practice of maintenance management, and to devise the best possible ways and means of organizing, planning, controlling and administering the maintenance function. Another direction is the development of theoretical maintenance models, which are practical and system-oriented. We have to recognize that considerable research efforts are required in order to close or at least reduce this gap. However, there exists wide potentials and a strong need for future research in the area of maintenance, both in theory and practice.

ACKNOWLEDGMENT

The author is indebted to Professor Rakesh Gupta of Adelphi University for his valuable suggestions made during the writing and review of this paper.

TABLE I

Percentage of Respondents Using Operations Research Techniques
in a Given Application Area

Application Area	Hovey and Wagner 1958	Schumacher and Smith 1964	Turban 1967	Ledbetter and Cox 1976
Scheduling	47%	90%	–	77%
Production Control	–	–	–	51
Project Planning/Control	–	–	–	55
Inventory Analysis/Control	45	90	–	64
Quality Control	33	51	–	37
Plant Layout	–	–	–	44
Equipment Replacement	15	27	–	30
Blending	–	–	–	47
Logistics (Transportation)	26	54	–	56
Plant Location	15	32	–	66
Forecasting	57	73	–	88
Advertising/Sales Research	20	27	–	69
Accounting	16	17	–	34
Capital Budgeting	11	39	–	71
Maintenance	16	32	16	22
Average Per Area	26	45	–	53

TABLE II

Utilization of OR Techniques in Various Production Applications

Application Area	N	Linear Programming	Simulation	Network Models	Regression Analysis	Queuing Theory	Dynamic Programming	Game Theory	Other
Ledbetter & Cox (1977)									
Production scheduling	56	53.6	46.4	10.7	8.9	16.1	12.5	0.0	17.9
Inventory analysis/control	47	23.4	57.4	6.4	25.5	8.5	6.4	2.1	14.9
Plant location	43	74.4	53.5	18.6	11.6	2.3	4.7	0.0	9.3
Logistics	41	65.9	58.5	19.5	14.6	7.3	2.4	4.9	4.9
Project planning/control	40	25.0	22.5	70.0	0.0	5.0	2.5	0.0	7.5
Production planning/control	37	51.4	48.6	18.9	8.1	10.8	8.1	0.0	8.1
Blending	34	34.1	17.6	2.9	8.8	2.9	0.0	0.0	2.9
Plany layout	32	40.6	59.4	15.6	6.3	15.6	0.0	3.1	9.4
Quality control	27	7.4	7.4	3.7	55.6	0.0	0.0	0.0	33.4
Equipment acquisition/replacement	22	18.2	50.0	4.5	0.0	4.5	0.0	0.0	31.8
Maintenance and repair	16	0.0	50.0	18.8	25.0	18.8	6.3	6.3	25.0
Turban (1967)	45	11.0	34.0	23.0	n.a.	7.0	2.0	n.a.	n.a.
Juul (1977)	n.a.*	n.a.	n.a.	25.0	n.a.	n.a.	n.a.	n.a.	n.a.

* n.a. - results are not available

(1)   Corder, A.S., <u>Maintenance Manage-ment Techniques</u>. McGraw-Hill, New York, 1976.

(2)   Davies, D.W., "Maintenance in Theory and Practice," <u>The Production Engineer</u>. September, 1976, pp. 463-465.

(3)   Fremont, G., "Maintenance Manage-ment: How's It Doing?," <u>Plant Engineering</u>. September 7, 1972, pp. 85-87.

(4)   Gnitecki, A. and Volkov, L., "Economical Efficiency of Mainte-nance Centralization and Speciali-zation in Industry," <u>Ekonomika Sovetskoj Ukrsiny</u>. No. 5, 1976, pp. 57-60 (Russian).

(5)   Hovey, R.W., and Wagner, M.M., "A Sample Survey of Industrial Oper-ations Research Activities," <u>Operations Research</u>. Vol. VI, No. 6, 1958, pp. 876-881.

(6)   Jardine, A.K.S., "The Use of Mathe-matical Models in Industrial Main-tenance," <u>The Institute of Mathe-matics and Its Applications</u>. August/September, 1976, pp. 232-235.

(7)   _____, <u>Maintenance Replacement and Reliability</u>. Wiley, New York, 1973.

(8)   Jordan, J.H., "Research in Contract Maintenance," <u>Techniques of Plant Engineering and Maintenance</u>. Vol. XIX, 1968, pp. 253-254.

(9)   Juul, P.T., "The Sad State of Main-tenance Management," <u>Plant Engin-eering</u>, February 7, 1977, pp. 125-128.

(10)   Konson, A.O., <u>Ekonomika Remonta Mashin</u>. Mashinostroenye, Leningrad, 1970.

(11)   Ledbetter, W.N. and Cox, J.F., "Operations Research in Production Management: An Investigation of Past and Present Utilization," <u>Production and Inventory Management</u>. Third Quarter, 1977, pp. 84-92.

(12)   Paulukonis, J., Staroselsky, V. and Chanin, M., "A Modern System of Instrument Repair in Lithuania," <u>Measurement Techniques</u>. 2, 1974, pp. 72-76 (Translated from Izmeritel naya Technika).

(13)   Posrednikov, N.E., "Ways to Increase the Efficiency of Equipment Mainte-nance," <u>Improvement of Industrial Production Efficiency</u>. Trkutsk (USSR), 1975, pp. 63-67 (in Russian).

(14)   Rohrmann, G.E., "Contract Mainte-nance in a Refinery," <u>Chemical Engineering Progress</u>. Vol. 67, No. 4, 1971, pp. 43-46.

(15)   Schumacher, Ch.C. and Smith, B.E., "A Sample Survey of Industrial Operations Research Activities 11," <u>Operations Research</u>. Vol. XV, No. 6, 1965, pp. 1023-1027.

(16)   Stuart, S.W., "Managing Maintenance: The State of Art," <u>Plant Engineering</u>. July, 1974, pp. 27-29.

(17)   Turban, L., "The Use of Mathematical Models in Plant Maintenance Decision Making," <u>Management Science</u>. Vol. 13, No. 6, January, 1967, pp. B342-358.

(18)   Woodward, Joan, <u>Industrial Organi-zation: Theory and Practice</u>. Oxford University Press, London, 1965.

CHAPTER 5

THE JAPANESE APPROACH TO MAINTENANCE

Reprinted from *Maintenance Management International* (formerly *Terotechnica*), Volume 2, 1981 pp. 79-86, Elsevier Scientific Publishing Company, P.O. Box 211, 1000AE, Amsterdam, The Netherlands

MAINTENANCE-ORIENTED MANAGEMENT VIA TOTAL PARTICIPATION

TOTAL PRODUCTIVE MAINTENANCE, A NEW TASK FOR PLANT MANAGERS IN JAPAN.

Yoshikazu Takahashi

Consultant, Japan Institute of Plant Engineers, Tokyo (Japan)

(Received April 26, 1980; accepted August 7, 1980)

Abstract

The historical development, philosophy and structure of Total Productive Maintenance (TPM), a uniquely Japanese approach to maintenance, are described.

INTRODUCTION

Total Productive Maintenance (TPM) is a particularly Japanese maintenance philosophy but it derived the greater part of its substance from a variety of non-Japanese mangement structures and practices, which were expertly adapted by the Japanese themselves, to the long-established Japanese traditions and culture.

1. WHY IS TPM OF SUCH IMPORTANCE TO JAPANESE PLANT MAINTENANCE?

Historical development

Productive Maintenance is a management tool learnt from the U.S. after World War II. A similar concept was, however, implemented in Japan after the Meiji Restoration of 1868, enabling her infant industry to adopt and adapt to the advanced industrial techniques then being utilised in Western countries. During this 100 year period, Japan's utmost effort was directed at industrialising herself in order to raise the gross national product, and the fullest utilisation of her industrial equipment was thus of primary importance.

However, it has been the last 25 years that have seen the most remarkable progress, in the scale and speed of continuous operation systems, in labour savings and in automation, realising a change in emphasis from manual work to machine operation. During this period, according to government statistics, the gross national product has multiplied by 30 times while the population itself has increased by only 30 per cent.

Until relatively recently, skilled workers and craftsmen in every part of the world were noted for cherishing their tools with the utmost care, improving them from day to day and even developing new ones to attain more effective means of production. Such endeavours and attitudes of mind are still praised under the name of "craftmanship". However, the recent trend in industrial activity has shown a drastic change in emphasis away from such tools and towards more complex plant and machinery, where it is of the utmost importance to understand such equipment, the function of each unit, the relationship between its condition and the quality of product, and the likely cause and frequency of failure of the critical items. Such understanding is directed at using the equipment to its fullest extent, for manufacturing to the desired quality and efficiency. Recognising this is the first step towards maintenance-oriented plant management, the advocacy of which is the burden of this paper.

Nature and composition of TPM

TPM is a method of achieving the above aim not via the easy route of capital investment but via a positive investment in human resources in order to fully utilise existing equipment through improved availability, through more assured quality and through labour-saving as a result of plant modification. This should be achieved at lowest possible cost, and accompanied by savings in energy and resources. Such efforts should result in a strong corporate policy, with which the company may withstand the difficult economic and political climate of the 80s. What is most needed here is the total participation of the work force, using the total systems approach in order to realise a total balance in efficiencies. In order to achieve this goal, it is evident that TPM implementation should be closely related to the profit plan and/or the cost reduction target of the company and that these should be thoroughly understood by managers at all organisational levels of the company.

The basic activities leading to company growth and profit are:

(1) research and development of those pro-ducts required by the customer, and also of the production plant required,

(2) reduction in the cost of such products,

(3) promotion of the sale of the products.

Development of a new product and of a new process may admittedly be achieved by a handful of talented engineers; history tells us that this has often been the case. However, in the area of production activity, where most employees are involved, it takes time for them to reach a stage where the whole range of equipment works so as to achieve the required production, assure the quality, and ensure the delivery time at reduced cost. The important point here, however, is the fact that these achievements require the support of each member of the organisation, from the first line workers to the plant management. With this in mind, the top management should recognise the need for applying TPM on a long-term basis to all equipment (including tools and dies, gauges and other instruments, etc.), so that the quality of product may be assured, equipment availability improved, delivery times met, environmental and safety hazards eliminated and finally, by means of these activities, equipment consciousness and a maintenance-oriented management established throughout the company.

Such a management climate will generate personnel who are expertly attuned to their equipment and to the plant as a whole. A policy of this kind, involving education and training courses, will make great demands on the time and perseverance of management.

2. WHAT IS THE BACKGROUND TO THE ADVOCATION OF TPM IN JAPAN?

The first motive: Improving productivity through a highly motivated workforce

Maintenance has long been regarded as non-rewarding and unglamorous in its nature. It was always the production department that was praised when a production target was met or the quality of product improved. Conversely, it was always the maintenance department that was blamed when equipment broke down. Maintenance was regarded as, at best, a necessary evil. However, this situation has drastically altered with the modern expansion in plant scale, complexity and sophistication and the maintenance function has been re-evaluated, particularly in capital-intensive industries, such as steel and chemicals, so that it now has an enlarged responsibility, authority and status. However, this in itself will not result in plant making its maximum contribution to company profitability. We have recognised here the importance of directing the attention of the workforce towards critical items of equipment.

During the 100 years since the Meiji Restoration, the Japanese educational system has registered remarkable progress. The Education Ministry has recently told us that the high school graduate (who has passed through 6 years in primary school, 3 years in junior high school and 3 years in senior high school; a total of 12 years) accounts for 94% of his age group. Japanese industry is therefore equipped with well educated personnel even at the first-line worker level. Naturally, ways have been sought to make best use of them. If they were employed on monotonous, repetitive, production-line work, what would become of their sense of work satisfaction or work accomplishment?

Now everyone knows that there are jobs that are monotonous and repetitive and that cannot be totally eliminated. In order to overcome this problem we have devised a programme of multicraft work or job enlargement in which all workers are given a range of challenging jobs in an effort to develop their skills at different crafts.

It is a feature of industrial maintenance that its technique and skills are relatively long-lived, and, therefore, once acquired, can continue to be of value as a medium for dis-

playing the technical expertise of the workforce. It therefore occurred to me that training programmes in this general area might be of particular value in helping to motivate our high school graduate workers. I therefore consulted with Seiichi Nakajima and other colleagues in the JIPE and ten years ago we launched a nationwide campaign of Total Productive Maintenance.

The second motive: The life-cycle approach to improving the overall performance of equipment

During almost 30 years of experience as a maintenance consultant I have frequently been frustrated by the gap or lack of information flow between those who project and design the equipment, those who manufacture it, those who maintain it and those who use it for production. This has impeded management on an equipment life-cycle basis and to overcome the problem it became evident that we should establish a systems approach to the measurement and evaluation of equipment performance at each stage of its life-cycle. The technical aspect of TPM emphasises Maintenance Prevention (MP) to meet this objective.

MP implies the design of highly reliable and easily maintainable equipment with subsequent feedback from maintenance and operation to design, and taking into due consideration such other factors as safety and operability. Operator familiarisation with the equipment and the promotion of a caring attitude to it will ensure good feedback of user information to the manufacturer. In this way we intend, via TPM, to establish a total systems approach to the equipment life-cycle (Fig. 1).

One year after the JIPE began its promotion of TPM, I participated in a conference organised by the American Institute of Plant Engineers (AIPE) in Anaheim California, on plant engineering and maintenance where a paper by Dennis Parkes suggested to us that terotechnology bore considerable resemblance

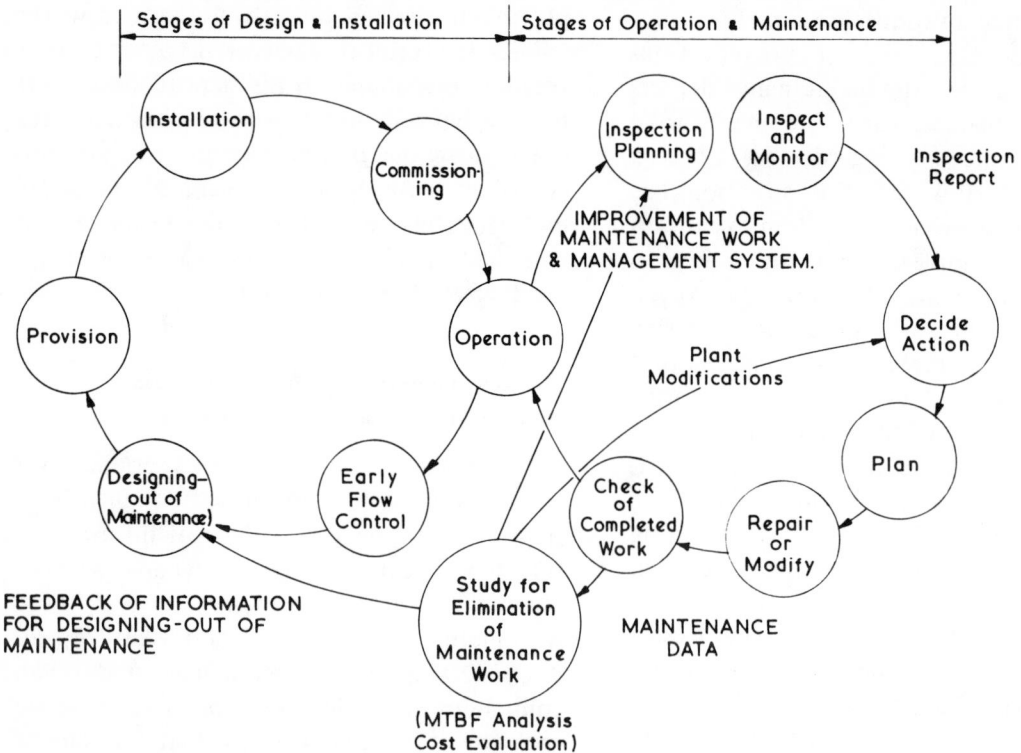

Fig. 1. Information flow for maintenance improvement.

to our TPM, the principal differences being firstly terotechnology's emphasis on economic evaluation of the whole life-cycle under the title "Life-Cycle Costing" (LCC), secondly the way in which its promotion in industry was positively supported by both government agencies and academic circles and finally in the participation, by manufacturers of equipment, in a collaborative effort with users, in the establishment of reliability and maintainability data banks. We therefore anticipated exchanging information on matters of mutual interest. Since then, the JIPE has sent annual study groups on this subject to the U.K. and to other European countries. In particular the group itenerary has, for several years, included a visit to the University of Manchester (who run an M.Sc. course in terotechnology) to attend a short programme of lectures and industrial visits directed by Tony Kelly.

The third motive: The voluntary small-group basis of TPM

It was in 1962 that the propagation of a small-group voluntary activity, original to Japan, began to be disseminated under the name "QC circle". The same year saw, in the United States, the start of a new movement called Zero Defect (ZD) (originating in the Martin Marietta Company in Florida) which was later introduced in Japan and combined with the existing QC circle to produce the kernel of a truly Japanised small-group voluntary activity called "ZD Group", generally aimed at the assurance of quality of products and of production processes. Since then this type of small-group voluntary activity has been widely practised in Japan regardless of the size of company. It was, therefore, natural that TPM followed the same promotional

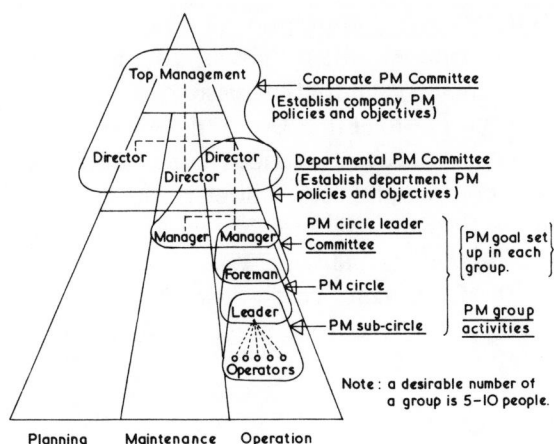

Fig. 2. TPM promotion system.

route in Japanese industry (Fig. 2).

Initially the introduction of TPM focusses small-group activity on the neatness (Seiri), tidiness (Seiton), cleanliness (Seiketsu), stainlessness, (Seiso) and orderliness (Shitsuke) of all physical assets. This "5S Activity" usually establishes itself as part of the routine work pattern at all levels of the company organisation. It is our opinion that it is not enough, for this purpose, merely to employ more maintenance men. There must also be a concentration on order and cleanliness, and on co-operative adherence to rules and regulations governing the working methods and therefore the quality maintenance work (Fig. 3). Quality of product or equipment availability can be jeopardised by a single dirty or greasy unit in a continuous flow process. The facilitation of maintenance actions will encourage operator equipment-consciousness; for example, the painting of crosses on critical bolts to indicate their correct position will result in ready identification of those which are loosening, obviating the necessity to test every bolt for tightness, a repetivive dehumanising task. Likewise, a thermo-tape affixed to a critical rotating part will readily indicate friction-induced heating. What is important here is to enable the operator to inspect and maintain with greater ease, which will also

help him to understand the equipment, particularly its critical parts. When he becomes readily responsive to his maintenance work, it can be said that he has reached the stage of assuring that his equipment is in its optimal condition. Another advantage of the simple visual check on equipment condition is that the operator can more readily call upon the co-operative wisdom and experience of his colleagues to assist him in his judgements.

Conclusion

To sum up, the conventional division of work and responsibility as exemplified in the phrases "I manufacture the equipment", "you maintain it" and "he uses it until it breaks down" will accomplish little and result in very slow progress. In the future, equipment, both ancillary (test instruments, repair tools, etc.) and primary productive, must be at the centre of our thoughts; how to improve its availability, how it can contribute to energy and resource conservation and hence to reduction in overall costs, and how this can be achieved through maintenance-oriented plant management as part of a total system approach. That is the basis of Total Productive Maintenance as it is at present being pursued in Japan, a general flow chart of which is shown in Fig. 4.

3. EXAMPLES OF THE ACHIEVEMENTS OF TPM

The monthly magazine "Factory Management" featured TPM in its April 1979 issue, giving excellent examples of TPM as demonstrated in 14 companies. All 14 companies are members of the Japan Institute of Plant Engineers and most of them have been awarded the "PM Distinguished Plant Prize". The following is a selection of these:
(1) Yokohama Rubber Co. Ltd., Mishima Factory. Product – automobile tyres. Number of employees – about 1,000.
  (i) Incidence of catastrophic failures – down 73%.

(ii) MTBF – five-fold increase.
(iii) Repair cost – down 47%.
(iv) Fuel consumption – down 40%.

(2) Chuo Spring Co. Ltd., Hekinan Factory. Product – automobile springs. Number of employees – about 400.

(i) Number of catastrophic failures – down 80%.
(ii) Defect rate during the manufacturing process – down from 1.5% to 0.5%.
(iii) Inventory/turnover ratio – down 50%.

(iv) Energy consumption – down 30%.
(v) Number of proposals for improvement – 10 fold increase.

(3) Fuji Photo Film Co. Ltd., Fujinomiya Factory. Product – various types of films. Number of employees – about 1,500.

(i) Rate of failure of equipment and facilities – down 72%.
(ii) Maintenance cost – down 23%.
(iii) Hours required for maintenance – down 50%.

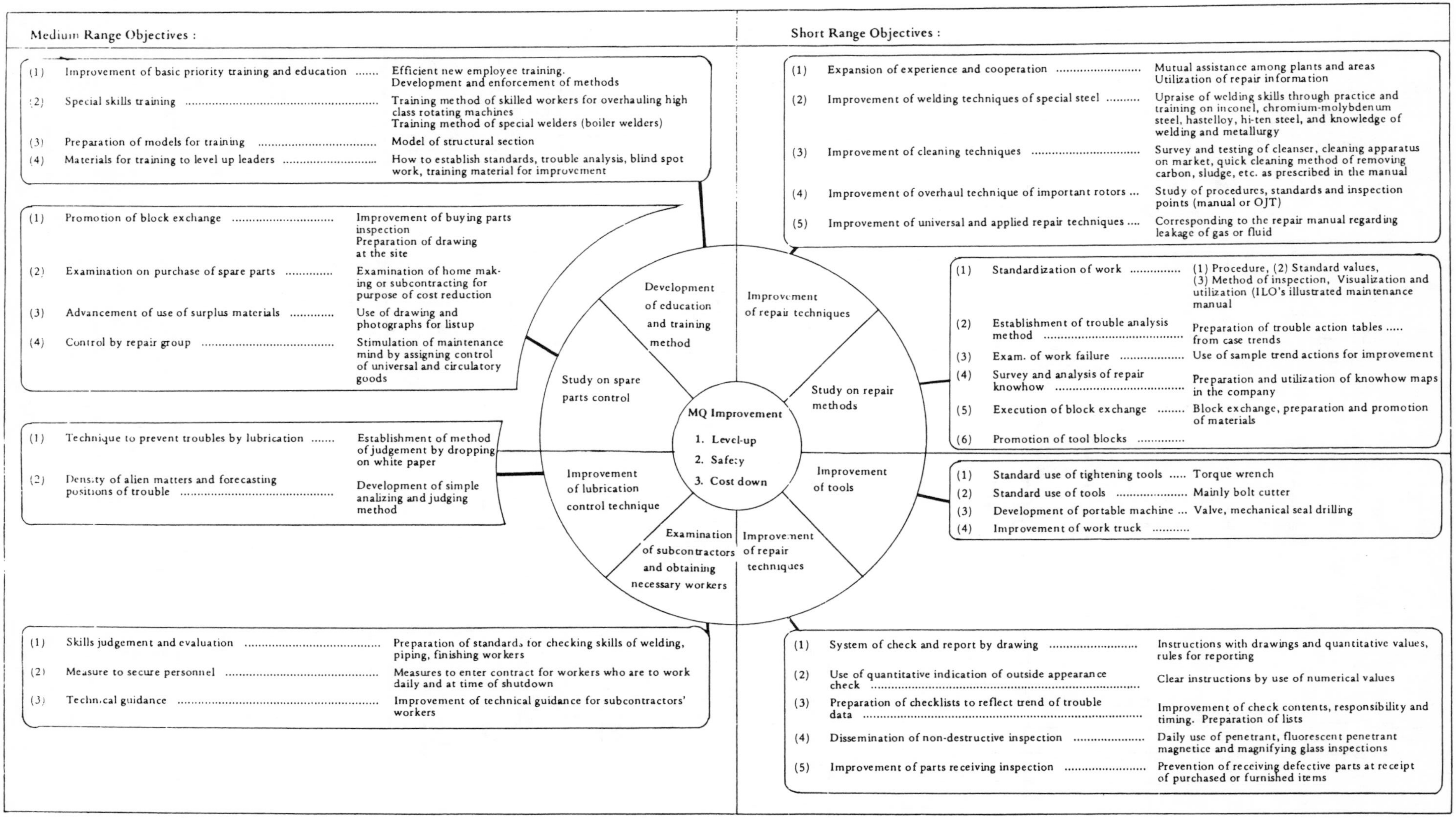

Short Range Objectives :

(1) Expansion of experience and cooperation ...... Mutual assistance among plants and areas / Utilization of repair information
(2) Improvement of welding techniques of special steel ...... Upraise of welding skills through practice and training on inconel, chromium-molybdenum steel, hastelloy, hi-ten steel, and knowledge of welding and metallurgy
(3) Improvement of cleaning techniques ...... Survey and testing of cleanser, cleaning apparatus on market, quick cleaning method of removing carbon, sludge, etc. as prescribed in the manual
(4) Improvement of overhaul technique of important rotors ...... Study of procedures, standards and inspection points (manual or OJT)
(5) Improvement of universal and applied repair techniques ...... Corresponding to the repair manual regarding leakage of gas or fluid

(1) Standardization of work ...... (1) Procedure, (2) Standard values, (3) Method of inspection, Visualization and utilization (ILO's illustrated maintenance manual
(2) Establishment of trouble analysis method ...... Preparation of trouble action tables .....
(3) Exam. of work failure ...... Use of sample trend actions for improvement from case trends
(4) Survey and analysis of repair knowhow ...... Preparation and utilization of knowhow maps in the company
(5) Execution of block exchange ...... Block exchange, preparation and promotion of materials
(6) Promotion of tool blocks ......

(1) Standard use of tightening tools ..... Torque wrench
(2) Standard use of tools ...... Mainly bolt cutter
(3) Development of portable machine ... Valve, mechanical seal drilling
(4) Improvement of work truck

(1) System of check and report by drawing ...... Instructions with drawings and quantitative values, rules for reporting
(2) Use of quantitative indication of outside appearance check ...... Clear instructions by use of numerical values
(3) Preparation of checklists to reflect trend of trouble data ...... Improvement of check contents, responsibility and timing. Preparation of lists
(4) Dissemination of non-destructive inspection ...... Daily use of penetrant, fluorescent penetrant magnetice and magnifying glass inspections
(5) Improvement of parts receiving inspection ...... Prevention of receiving defective parts at receipt of purchased or furnished items

Medium Range Objectives :

(1) Improvement of basic priority training and education ...... Efficient new employee training. Development and enforcement of methods
(2) Special skills training ...... Training method of skilled workers for overhauling high class rotating machines / Training method of special welders (boiler welders)
(3) Preparation of models for training ...... Model of structural section
(4) Materials for training to level up leaders ...... How to establish standards, trouble analysis, blind spot work, training material for improvement

(1) Promotion of block exchange ...... Improvement of buying parts inspection / Preparation of drawing at the site
(2) Examination on purchase of spare parts ...... Examination of home making or subcontracting for purpose of cost reduction
(3) Advancement of use of surplus materials ...... Use of drawing and photographs for listup
(4) Control by repair group ...... Stimulation of maintenance mind by assigning control of universal and circulatory goods

(1) Technique to prevent troubles by lubrication ...... Establishment of method of judgement by dropping on white paper
(2) Density of alien matters and forecasting positions of trouble ...... Development of simple analizing and judging method

(1) Skills judgement and evaluation ...... Preparation of standards, for checking skills of welding, piping, finishing workers
(2) Measure to secure personnel ...... Measures to enter contract for workers who are to work daily and at time of shutdown
(3) Technical guidance ...... Improvement of technical guidance for subcontractors' workers

Center diagram:

MQ Improvement
1. Level-up
2. Safety
3. Cost down

- Improvement of repair techniques
- Study on repair methods
- Improvement of tools
- Improvement of repair techniques
- Development of education and training method
- Examination of subcontractors and obtaining necessary workers
- Study on spare parts control
- Improvement of lubrication control technique

Fig. 3. A system for maintenance quality (MQ) improvement.

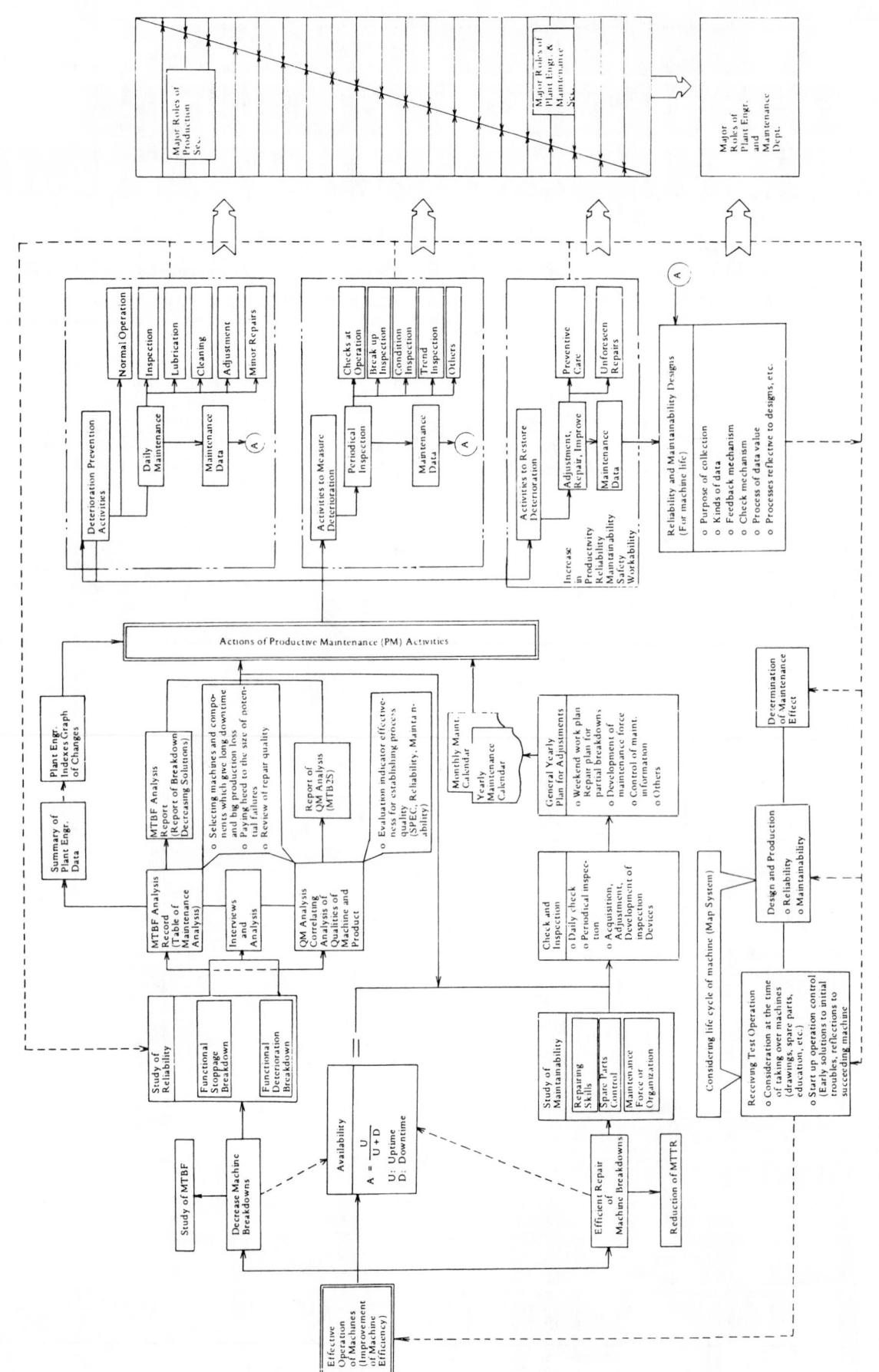

Fig. 4. General flow chart of TPM activities (QM Analysis is Quality-Machine Analysis).

Reprinted by SME from *Iron Age*

DETROIT'S AUTO INDUSTRY WRESTLES WITH MACHINE DOWNTIME

Worker morale and plant spending have suffered by the general lack of concern for effective preventive maintenance programs.

By Bryan H. Berry

Detroit's failure to maintain its machine tools has sunk automotive uptime on many operations to 50-60 pct averages and exacted costly tolls in plant spending and worker morale. Detroit's automotive industry has been so busy pushing out product by round-the-clock production that it has spurned maintenance. But current hard times are waking up automakers to the need for preventive maintenance.

"We don't have maintenance managers," says James W. Kolp, superintendent, plant engineering, at Buick Motor Division in Flint. "We have crisis managers. If the product lines are selling well, machines have to run extended hours to meet production schedules, so you limit your maintenance time access to the equipment. On the other hand, if the product isn't selling well, you don't generate money, so you don't work your people any extra hours to take care of the maintenance.

"Yet we generally still seem to work our maintenance people seven days a week. If you don't do good maintenance your equipment condition eventually de-generates to such an extent that your productivity drops off. Then you end up working premium time to make production that you should be getting on straight time."

The maintenance operator is dispatched like a fireman from fire to fire. Predicting the needs of machines maintained on no recorded schedule is a guessing game, so breakdowns seem random. The maintenance operator becomes dispirited daisychaining from crisis to crisis, often returning to the same machines he hastily repaired earlier. Since the maintenance staff is hired to handle such peak periods of breakdown, many sit idle during the slow down. Job satisfaction at both extremes

261

is minimal. (Such labor issues, including one that many managers consider a significant contributor to downtime—the lines of demarcation between the skilled trades—will be the subject of an *Iron Age* article in the Sept. 28 issue.)

Uptime throughout the automotive and supplier industries including machining and assembly plants and foundries averages 50-60 pct, according to sources in both industries. "If automakers get 60 to 65 pct on an in-line transfer machine, they're delighted," says Philip J. Pressler, regional manager in the Detroit area for Kingsbury Machine Tool Corp.

Such low uptime is expensive. It means more capital expenditure. Plants overfacilitize as much as 40-50 pct in order to back up malfunctioning equipment. "Automakers will say, 'We need two machines instead of one,'" says Mr. Pressler. "In some cases we will say, 'No, you don't. You only need one machine if you run it properly.' They say, 'That means you'll sell one less machine.' We say, 'If you don't get your productivity up, we won't sell any machines!'"

Automakers overspecify their equipment orders, requesting, for example, load detectors on spindles to warn excessive dullness rather than anticipating it by regular tool changes. Automakers expect equipment breakdowns, so they stockpile parts to feed the line while the machine is being repaired. High-technology equipment, such as electron beam welders, are frequent troublemakers; automakers average 50 pct uptime with them.

Not all automotive operations have such low uptimes. Some get up into the high 80s, and one transfer line achieves 90 pct uptime. The machine tool builder attributes that exceptional figure to 308 aluminum. "You can't lose sight of the material of the machine," he says. "308 aluminum will last much longer than cast iron, so you can achieve greater efficiencies with it. We have one plant where they actually are monitoring the cast iron to make sure it doesn't get too hard."

The 90 pct uptime may be an exception in the United States, but it is an average in Japan. One American machine tool builder tells of selling his machine tools to both a Japanese and American automaker to run exactly the same parts. The tool ordered by the Japanese had less diagnostic equipment and cost half that of the one for the American. The Japanese automaker runs it two shifts at 90 pct efficiency, while the American runs it

Average Downtime for all Ford Operations

Including transfer lines, stand-alone equipment,
24-hour glass and plastics operations: 23 pct

The 23 pct downtime is comprised of:

11 Pct Machine Failure
▶ 60 pct people-oriented problems such as lines of demarcation
▶ 40 pct purely machine problems

8 Pct Inefficient Arrangement of Lines
▶ lines too long
▶ piece stockpiles between machines not managed properly
▶ too many spindles at stations

4 Pct Problems with Workforce
▶ includes absenteeism and long breaks

Data is called from 3,500 different studies of Ford operations, 1974-80

Source: Ford North American Productivity Office

Data gives actual pct of performance compared to machine optimum for 154 transfer systems used throughout Ford Motor Co., 1974-80.

Transfer Line Uptime: 46-64%

	Totally Synchronous	Partially Non-Synchronous	Totally Non-Synchronous
	Pct	Pct	Pct
Major Systems (100-300 Stations)	46	51	64
Intermediate Systems (50-99 Stations)	61	55	57
Minor Systems (10-49 Stations)	52	56	59
Total Systems	53	56	61

Note: Synchronizing Stations Cuts Uptime

three shifts at 52 pct. At a factor of 100, the Japanese scores 180, the American 156. Consider that the American is payrolling labor for eight hours more than the Japanese, and you begin to see the enormity of the transPacific disparity. Automakers here are starting to hear the message loud and clear: Preventive maintenance works there and is needed here.

Preventive maintenance, of course, is not new to these shores and was not invented in Japan. It's just that Americans forgot it when production ballooned; the Japanese stuck with it.

"We plan to copy from Japan whenever we can," say Herbert L. Misch, Ford's vice president, environmental and safety engineering.

"There's no place for the not-invented-here attitude."

Ford decided to investigate Japanese manufacturing techniques when the company ordered some automatic transmissions from that country and "they beat us soundly in every factor of the program," says Robert J. Burger, manager, chassis manufacturing and industrial engineering departments, Ford transmission and chassis division.

"We are the transmission experts, yet these people who have never built an automatic came in and did it for something like half the cost." Traffic by Ford personnel to Japan has been steady since. What are they learning, and what are the American automakers doing to raise uptime?

One key difference between the two countries is that the Japanese run their machines slower than we do—at 40-second cycles versus our 18-20-second cycles, estimates Mr. Burger. Slower lines require fewer workers. Faster machines have a greater chance of accident and give the operator less time to position the part correctly.

"The so-called fast transfers are nice when they work," says Donald W. Carlson, manager, manufacturing engineering dept., Ford engine division, "but if you ever get a block katywampus in that thing, it crashes, it's just unbelievable."

"In one wreck on those fast transfers you lose everything you have gained," says Gerry C. Charbeneau, supervisor, maintenance control section, maintenance and environmental control dept., Ford transmission and chassis division. "You are down for hours—and days."

"The Japanese are slow and steady," says Mr. Carlson. "We are fast and erratic."

Americans have more machining stations on their transfer lines than the Japanese do. The greater number of stations increases the chance of breakdown; if any one station goes down, the whole line goes down. Sometimes you can bypass the malfunctioning station, but not if it's a critical one.

Americans are learning to build a greater number of independent sections with fewer stations on each.

"Splitting up the machines into different sections and electronics improvements have caused the biggest gains in uptime we have made," says Frank J. Starr, director of manufacturing engineering at GM Hydraumatic's Ypsilanti (Mich.) plant.

Americans expect defective parts and plan for repair. "We assume in our planning 3-5 pct repair," says Mr. Burger. "We have enough repair equipment and enough capacity to cover that 3-5 pct loss for repairs and out of the repairs comes 2 pct scrap. We don't plan to make it, but history has shown that we will.

"When something goes wrong with a transmission in Japan, the operator takes it off the line, sets it on the floor, continues his work on the line, and before he goes home he repairs that transmission and then puts it back on the line. There is no loss. The transmission doesn't go into a big repair area where hordes of people try to fix it. He does his own repair." The Japanese refusal to factor repair into their planning has a way of diminishing its effect on their plans.

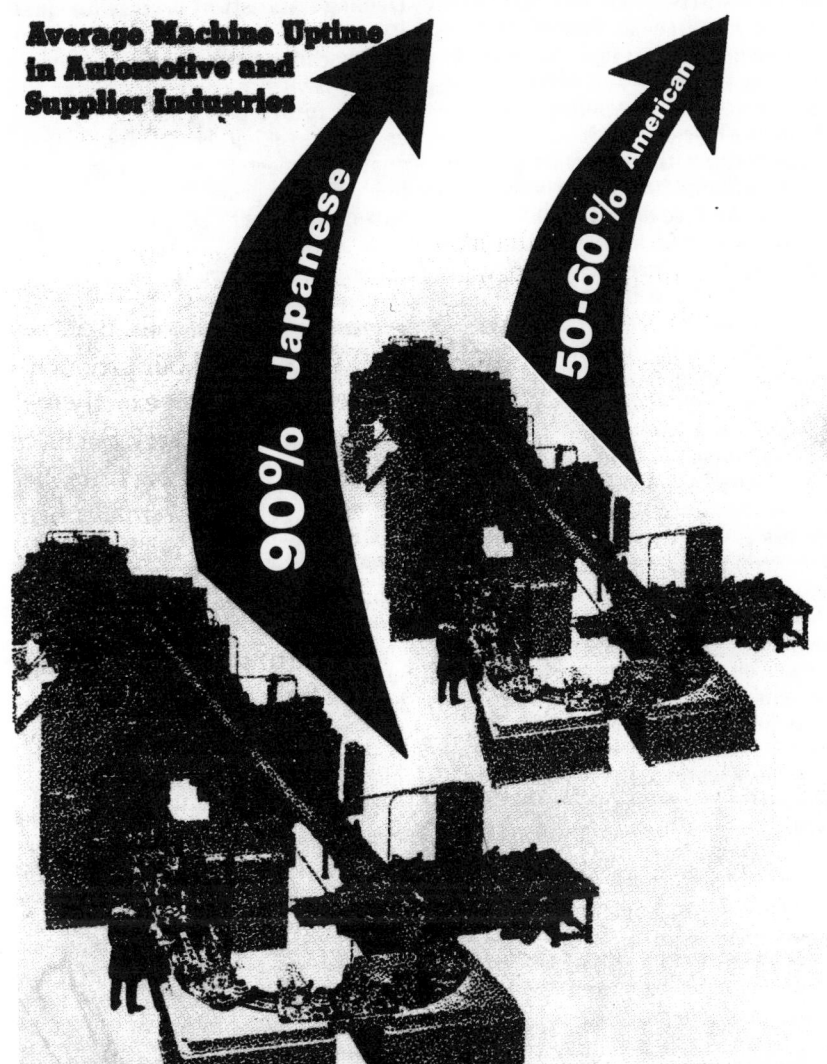

A leading American machine tool maker sold a transfer line to both an American and a Japanese automaker to run identical parts. The line delivered to the Japanese customer was less sophisticated and cost half that of the one delivered to the American. The Japanese is running the machine two shifts and getting an uptime of 90 pct. The American is running three shifts and getting 52 pct uptime. The figures are typical of the Japanese and American auto industries, say manufacturing personnel in the automotive and machine tool industries. American automakers average 50-60 pct uptime and are happy to exceed 65 pct.

Similarly, the Japanese do not stockpile—bank, or float, as it is termed—parts to feed the line when a machine goes down. "The Japanese have no excess inventory to cover bottlenecks," says Mr. Burger. "If a bottleneck operation goes down, everybody covers it because there are not enough pieces to keep subsequent operations going, so you've to get the machine back up.

"What we do is run the hell out of bottleneck operations to get enough float so that when it goes down we will have time to work on it. That time is what kills us. We know we have two hours' float, so we take two hours to bring that machine back up. Work has a way of filling the hours allocated to it, as Parkinson said. The

Japanese system says: 'There is no time. You've got to get in there and fix it. Everything keeps running.'"

Both the major difference between the Japanese and American maintenance strategies is the Japanese insistence on prevention. "The emphasis of Japanese management," says Tetsuo Koyama of the Japan Productivity Center in Washington, D.C., "is on keeping the day-to-day preventive maintenance on a level high enough that all major overhauls can be scheduled for between shifts. Maintenance not concerned with the moving line is done by the line worker on a more-or-less continuous basis."

"The Japanese program all their major tool changes for the off shift or between shifts," says Mr. Burger, "whereas our tool changes are done in our run shifts. Tool changes on a transfer machine can account for two or three hours of an eight-hour shift." The transmission and chassis division, he says, is rethinking the value of round-the-clock shifts and in fact is considering the economic feasibility of running 80 hours a week, two shifts a day, five days a week.

"Sixty pct of the Japanese maintenance labor is preventive," says Mr. Charbeneau, "and it is on only 10 pct of their assets, the critical ones." Very little of current American maintenance is preventive.

"The auto companies are unwilling to spend the money for preventive maintenance," says Irving J. Bluestone, the UAW's vice president for GM affairs until last year and now university professor of labor studies at Detroit's Wayne State University. "They let the equipment go until it breaks down, and then when it breaks down, they've got to do these rush repairs. You put it together with spit and chewing gum and let it run a little longer before it breaks down. There are very few companies in my experience that actually budget for preventive maintenance. They wait until something happens—like hospitalization."

Now, however, Ford and General Motors are instituting programs to plan preventive maintenance.

Both companies gradually will reassign workers presently on breakdown maintenance to preventive maintenance. The maintenance personnel will enter on computers information about past tool changes and machine maintenance and performance; these data will enable them to predict future performance and maintenance needs.

"Our management information system," says Mr. Kolp, chairman of the planned maintenance subgroup of GM's corporate maintenance task force, which currently is mapping GM's maintenance strategy, "will start gathering information about life cycle change of components, which is determined by the manufacturer, the environment the equipment is running in, experience, and when a bearing, a slide, a motor, a valve, a cylinder needs to be replaced. At the same time maintenance people will enter equipment downtime. You can start getting a trend analysis on cycle time." Maintenance personnel will study the different results from the same equipment and determine the maximum cycle for the equipment and how best to get it.

Ford is implementing a similar

program based on Dr. W. Edwards Deming's statistical process (or quality) control. Machine operators monitor machine performance to ensure that it is within process limits and to predict future performance. No longer will checking be left entirely to quality inspectors at the end of the line who spot problems only when the out-of-spec parts reveal out-of-spec tools up the line.

Instead, the operator will be able to anticipate problems. The operator

Ford now plans to do block tool changes and major preventive maintenance on the third shift.

records on a graph-paper table the changing dimensions of the tool. "The operator can look at his first few checkpoints on the table," says Mr. Carlson, "and get an idea of where the curve will go."

In addition, Ford recently began a computer-based block tool change program to coordinate tool changes primarily on transfer machines and machines, such as broaches, with long tool cycles. At the Van Dyke (Sterling Heights, Mich.) chassis plant later this year, Ford will combine the computerized block tool change program and a preventive maintenance program; Ford plans to institute similar programs throughout the company.

"We find in the inline transfer machine," says Mr. Burger, "a number of tools with the same cycle time, and we change them in a block. At the same time we shut off the machine to change the block of tools, the foreman has a computer printout sheet that tells job setters to go to these oil points, check these oil levels, check that limit switch position.

"Heretofore the machines would go down for tool changes and go back up and five minutes later come back down for maintenance. This marriage of maintenance and tool change is so simple, I don't know why it hasn't been done before."

"When people talk about tools," says Mr. Carlson, "they think of drills and taps. But there are other things in a machine that really behave like tools. When you look at the locators, spindle bearings, limit switches, and fixture details, those are really tools. The only difference is that their frequency of change is much longer. Anything with a finite life in that machine really has to be looked at

like a tool."

Lubrication also is key to preventive maintenance. "Keeping everything lubricated may solve half your maintenance problems," says Mr. Kolp.

Ford and General Motors will shift maintenance workers from crisis to preventive maintenance on a gradual basis.

Limiting major maintenance to off or split shifts is key to the high uptime of Japanese production runs. They run the two production shifts back to back and then do major maintenance during the next six or eight hours (off shift); or they wedge two- or three-hour maintenance periods between both shifts.

As the American automakers switch from crisis to preventive maintenance, will they too be able to limit major maintenance to an off shift or split shift so as not to rob production time?

"I don't know," says Mr. Kolp. "When I was department head of the foundry, we used to do it that way. The first and second shifts, we did inspection maintenance as part of the regular planned maintenance—such things as vibration checks and lubrication. We provided immediate service for any breakdowns at that time plus we built up modules for the third shift. The third shift we did all the repairs that we had identified earlier plus what we knew needed repair from our past experience, records, and the life cycle changes.

"I suspect that as GM grows into its planned maintenance program, a lot of this kind of work will be done on the third shift. That is something we will evolve into depending on what each plant needs."

Ford now plans to do block tool changes and major preventive maintenance on the third shift, since low production demands don't require a third shift anyway these days. "But if production requirements ever demand that third shift again," says Mr. Burger, "I don't know what the heck we are going to do. We need to get our data base in order to make sure we can achieve in two shifts what we used to do in three shifts." Ford might have to scuttle parts of the maintenance and tool change programs until it has enough equipment to meet production demands in two shifts. That would require a considerable investment. □

Reprinted from *Automotive Industries*, July 1982

MAKING TWO SHIFT PRODUCTION PAYOFF

By changing from three shifts to two shifts and devoting that open time to planned maintenance, many automotive operations have the potential for improving quality and productivity, while saving a million dollars a month.

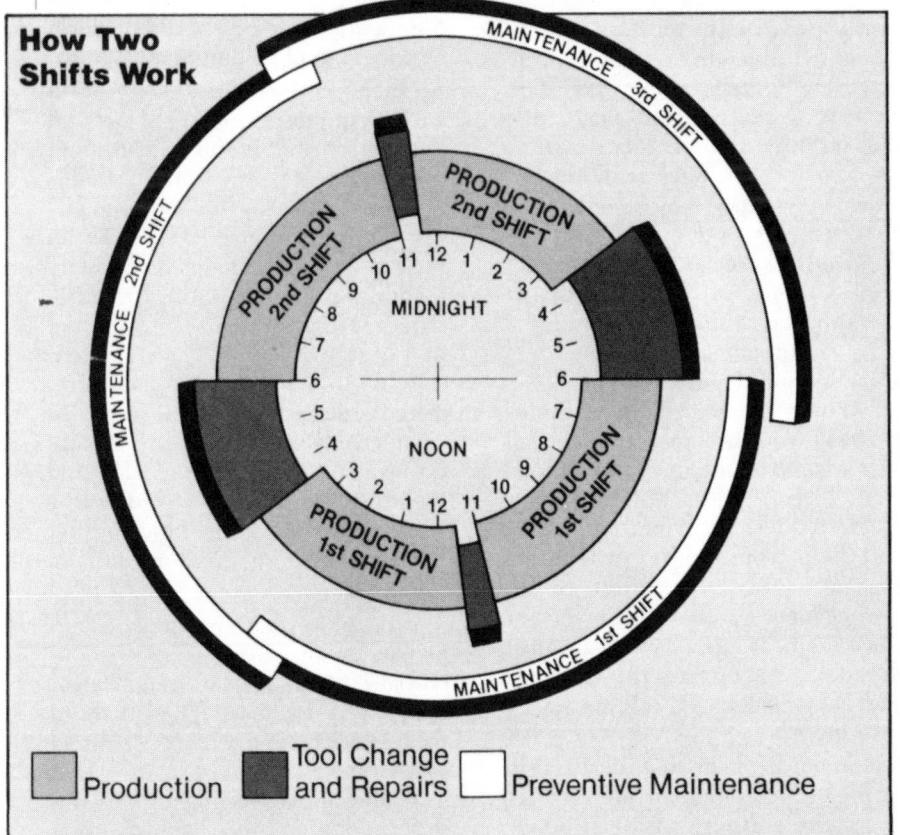

How Two Shifts Work

□ Production   ■ Tool Change and Repairs   □ Preventive Maintenance

The production shift at Chevy's Flint Motor plant starts at 6 am and runs to 3:30 pm, with a half hour for lunch at 11 am. During lunch, simple tool changes and basic preventive maintenance take place. Major tool changes and repairs take place during the 2½ hours after the first shift. The same process is repeated for the second shift. Note that the maintenance crew's three shifts overlap each other.

by John McElroy

One of the greatest roadblocks to US machining operation productivity today is downtime. Average auto industry transfer line uptime hovers around 55%, meaning those lines are sitting idle nearly half the time they are scheduled to be running.

But a startling change is beginning to take place in several automotive factories. By taking advantage of current low volumes, they are changing from three shifts per day to two shifts, and devoting that open time to preventive maintenance. Surprisingly, they are producing about the same amount of parts as they did before and productivity has shot up. Consider these examples:

● At Chevrolet's Flint Motor plant, scheduled production on the block line was changed from three to two shifts. Productivity jumped nearly 50% and the plant is saving $1.5 million annually.

● At Buick's engine plant, production on the block line was changed to two shifts. Productivity gains suggest this action will free up a whole block line which may be retooled for a new engine.

● Saginaw Steering has switched to two shifts on two transfer lines that make housings for hydraulic steering pumps and steering gears. Uptime improvements of 10% to 20% are expected.

● At Ford, all machining operations have been reduced by 10 to 30 hours per week. At most, only 5% production capacity has been lost, while costs are way down and productivity is way up.

How is it possible that an entire shift can be taken out of the system while output remains virtually the same? James Harbour of Harbour & Associates, a manufacturing consulting firm, explains.

"Nobody builds an engine faster than anyone else," he says, also noting that the same applies for transmissions and axles. "They all run the same feeds and speeds on the same type of equipment that they've bought from the same machine tool builders. So the question to ask is not, who has the best gross

(Continued)

output per hour, it is who has the best net output per week?"

The key to getting more net output per week, says Harbour and other manufacturing experts, is to prevent the production system from breaking down when it is scheduled to run, by doing more preventive maintenance.

"When you run three shifts, there's always a struggle between production and maintenance," explains Joe Dalton, supervisor of industrial engineering of manufacturing staff at Ford. "You either cut into one or the other."

Paul Guy, Ford's director of manufacturing engineering staff, explains why most US machining operations run on three shifts. "For many years we operated under the philosophy of high asset utilization to maximize output from existing facilities and to avoid building additional plants.

"From an investment standpoint that makes good sense, but you don't get your scheduled production because of unplanned downtime. So why not schedule your downtime and control it? That's the new theory."

James W. Kolp, general works engineer at Buick, explains it even more succinctly. "You get into this funny thing called getting production out the door," he says. "This kind of logic says that if you're fixing something, then it's not making parts."

It should be noted that two shift operations are not feasible in certain production systems such as foundries, injection molding, heat treatment and diecasting where equipment can't be left to cool down. Assembly operations, which by their very nature must be maintained daily to insure quality and run at high rates, have always been on two shifts. But three shifts in machining operations can wreak havoc on the equipment, causing over three hours of downtime per day in breakdowns and tool changes.

By changing from three to two shifts, the extra time can be alloted for maintenance and tool change so that production doesn't have to stop. The best way to do this is to split the two shifts instead of having one right after the other.

TWO SHIFT PRODUCTION

Chevrolet's Flint Motor plant uses two split shifts on the 1.8-L (110-in.^{3}) four-cylinder block lines. Each shift operates for 9 hours with 2½ hours between shifts. Minor tool changes and preventive maintenance are done midway through each shift during the half hour for lunch when the machinery is not running.

Major tool changes and repairs are done during the 2½ hours between shifts. Maintenance crews overlap on both shifts so that there is a smooth transition on any repair work. Now the plant produces nearly 100 blocks an hour whereas the average General Motors engine plant only produces 52 an hour.

Preventive Maintenance Payoff

The increased uptime that results from preventive maintenance not only smooths and balances out operations, it reduces scrap and increases quality. In addition, it allows operations to reduce in-process inventory that is kept on hand to keep the lines running when a machine breaks down unexpectedly.

In many three-shift operations transfer lines are divided into different sections with large banks in between sections. And the motivation to repair the breakdown seems to run in inverse proportion to the amount of inventory on hand. In remarkable contrast, one superintendant at a transmission plant that runs on three shifts boasts that they can run for two days with the in-process inventory they have banked on conveyors in the roof, whereas the two-shift block line at Chevrolet Flint keeps less than one shift of blocks or rough stock in inventory.

Fewer unplanned breakdowns are also easier on the machinery.

"When you do planned maintenance, you become what I call being smarter than your equipment," says Buick's Kolp. "Take a speed reducer. A vibration check will help you predict when the bearings will fail. If you replace the bearings before they fail, you replace the bearings. If you wait until they fail, you replace the bearings and the gear-set. You tell me which costs more."

Harbour explains that most transfer lines for engine blocks are tooled to produce a gross capacity of about 135 an hour, but most three-shift operations only achive a net of 75 an hour and many are well below that. In 24 hours, therefore, a transfer line is theoretically capable of producing 3240 blocks, but in fact only achieves a net of 1800 blocks or only 55% of its potential.

On a split two-shift operation with eight hour shifts, says Harbour, the gross output is still 135 an hour, but with extra preventive maintenance increasing the uptime, the net output rises to 100 blocks an hour. So, he says, the gross capacity for one day's operation is 2160 blocks (135 x 16 hours) and the net output is 1600, or 74% of its potential. In other words, it is possible to achieve almost a 20% increase in productivity for absolutely no capital expenditure. Harbour adds that with an extra hour per shift, or by working Saturday, the total output perweek will be greater than an around-the-clock operation.

But what has really grabbed management's attention is the potential cost savings. Harbour calculates that a typical engine made in the US contains 6.8 hours of direct and indirect labor, but he says that if the entire plant changed to a two-shift operation, the labor content would drop to 5.2 hours. He then multiplies this 1.6 hour difference by the average automotive wage of $20 an hour. This savings, he notes, works out to $32 per engine. And since a typical engine plant will build over 8000 engines a week, the savings can run over $250,000 a week!

Ford has decreed, through a directive signed by company president Donald Petersen, that no machining operations may run more than 123 hours per week, down from 133 hours. The directive goes on to say that a two-shift operating plan must be considered as an alternative to all three-shift operations, whenever possible.

"I'll guarantee you," says Ford's Paul Guy, "there won't be a new machining plant built [at Ford] without being planned for a split two-shift operation."

Says Buick's Kolp, "You have to sit down with the hourly and production people and do a sell job on it and show its benefits, because it changes the routines in their lives."

Goals For The Future

Split two-shift operations are a new phenomenon in US automotive operations, but they are the rule of thumb in Japan where some transfer lines achieve 90% uptime. Buick's engine plant now hopes to achieve 90% uptime within five years.

Buick's Kolp suggests that they may even have to go one step further. "Maybe there shouldn't be any production people or tool changers in a highly automated system," he says. "Maybe they all ought to be just skilled trades people who are responsible for production output and at keeping the equipment running. That's a pretty big step for anyone to swallow, but that's how radical we're thinking in some cases." ∎

Output Jumps With Two Shifts
Chevy Flint Motor Plant #5—Block Line

	Line #1 Milling	Line #2 Rough Turning	Line #3 Finishing
March 1981 Three Shifts + Sat.	70 parts/hour	66 parts/hour	64 parts/hour
March 1982 Two Shifts (9 hrs.)	104 parts/hour	97 parts/hour	95 parts/hour
Improvement	**49%**	**47%**	**48%**

INDEX

C

Owners, 146
Oxidation, 22

P